Cambridge
International AS and A Level Mathematics

Pure Mathematics 1
Practice Book

Greg Port

HODDER
EDUCATION
AN HACHETTE UK COMPANY

Answers to all of the questions in this title can be found at www.hoddereducation.com/cambridgeextras

Questions from the Cambridge International AS and A Level Mathematics papers are reproduced by permission of Cambridge International Examinations.

Cambridge International Examinations bears no responsibility for the example answers to questions taken from its past question papers which are contained in this publication.

This text has not been through the Cambridge endorsement process.

Every effort has been made to trace all copyright holders, but if any have been inadvertently overlooked the Publishers will be pleased to make the necessary arrangements at the first opportunity.

Although every effort has been made to ensure that website addresses are correct at time of going to press, Hodder Education cannot be held responsible for the content of any website mentioned in this book. It is sometimes possible to find a relocated web page by typing in the address of the home page for a website in the URL window of your browser.

Hachette UK's policy is to use papers that are natural, renewable and recyclable products and made from wood grown in sustainable forests. The logging and manufacturing processes are expected to conform to the environmental regulations of the country of origin.

Orders: please contact Bookpoint Ltd, 130 Milton Park, Abingdon, Oxon OX14 4SB. Telephone: (44) 01235 827720. Fax: (44) 01235 400454. Lines are open 9.00–5.00, Monday to Saturday, with a 24-hour message answering service. Visit our website at www.hoddereducation.com.

© Greg Port 2013

First published in 2013 by
Hodder Education, an Hachette UK Company,
338 Euston Road
London NW1 3BH

Impression number   5 4 3 2 1
Year   2017 2016 2015 2014 2013

Cover photo © Joy Fera/Fotolia
Illustrations by Datapage (India) Pvt. Ltd
Typeset in 10.5/14pt Minion Pro Regular by Datapage (India) Pvt. Ltd
Printed in Great Britain by CPI Group (UK) Ltd, Croydon, CR0 4YY

A catalogue record for this title is available from the British Library

ISBN 978 1444 19633 7

Contents

1 Algebra

Background algebra, Linear equations, Changing the subject of a formula

1 Simplify these expressions as fully as possible.

(i) $2(a - 3b) - 3(b - 3a)$

(ii) $7cd(d^2 - 2) - 3cd^2(8d + 5c^3)$

(iii) $7a + 3b \times a - 4a^2b \div 2ab$

(iv) $\dfrac{8f^4}{3g} \times \dfrac{9g^3}{12fg}$

(v) $\dfrac{16y}{3x^2} \div \dfrac{8y^2}{9x^3}$

(vi) $\dfrac{24 - 16x}{3x - 2x^2}$

(vii) $\dfrac{4}{3x} + \dfrac{3}{4x} + \dfrac{2}{5x}$

(viii) $\dfrac{x-1}{4} + \dfrac{5-x}{5}$

2 Factorise fully.

   (i) $12mn^2 + 9mn^3$ **(ii)** $p^2 - p - 12$

   (iii) $3q^2 + 5q - 2$ **(iv)** $ts + tp - 2us - 2up$

3 Solve for x.

   (i) $2(x + 5) = x - 7$ **(ii)** $\frac{1}{2}(6x + 8) - 3 = 9 - \frac{3}{2}(4 - 10x)$

4 Hakim drives from Auckland to Hamilton in 2 hours.

Ravi leaves at the same time as Hakim and drives the same route at, on average, 4 km/h slower and arrives 6 minutes after Hakim.

Find the distance from Auckland to Hamilton.

5 Make the letter in brackets the subject of the formula.

(i) $\dfrac{v}{b} - c = \dfrac{d}{e}$ *(e)*

(ii) $km^2 + n = p - wk$ *(k)*

(iii) $\dfrac{1}{f} = \dfrac{1}{u} + \dfrac{1}{v}$ *(v)*

(iv) $\sqrt{d - 3e} = \dfrac{1}{2\pi}\sqrt{\dfrac{p}{w}}$ *(e)*

Quadratic equations, Solving quadratic equations

1 Solve these quadratic equations.

   (i) $x^2 + 5x = 0$ **(ii)** $x^2 - 25 = 0$

   (iii) $x^2 - 2x - 8 = 0$ **(iv)** $x^2 + 5x - 14 = 0$

   (v) $x^2 - 3x - 40 = 0$ **(vi)** $2x^2 - 5x - 3 = 0$

(vii) $2x^2 - x - 3 = 0$

(viii) $3x^2 - 5x - 2 = 0$

(ix) $5x^2 + 13x - 6 = 0$

(x) $3x^2 - 6x + 3 = 0$

(xi) $9x^2 - 1 = 0$

(xii) $6x^2 + 7x - 3 = 0$

(xiii) $3x^2 - 6x = 0$

(xiv) $12 = 18x^2 + 15x$

2 Solve the following equations.

(i) $x^4 + 3x^2 - 4 = 0$

(ii) $5 - \dfrac{2}{x} = 2x$

(iii) $x + 2\sqrt{x} = 8$

(iv) $x^6 + 8 = 9x^3$

(v) $\dfrac{3}{x^4} - \dfrac{11}{x^2} = 4$

(vi) $\dfrac{2}{x} - 1 = 4 - \dfrac{9}{\sqrt{x}}$

Equations that cannot be factorised, The graphs of quadratic functions

EXERCISE 1.3

1 Write these quadratic expressions in completed square form $(x \pm a)^2 \pm b$.

(i) $x^2 - 6x + 1$

(ii) $x^2 + 4x$

(iii) $x^2 - 3x + 2$

(iv) $x^2 + 2x + 5$

2 Using your answers to question 1, solve the following equations.

(i) $x^2 - 6x + 1 = 0$

(ii) $x^2 + 4x = 0$

(iii) $x^2 - 3x + 2 = 0$

3 Write these quadratic expressions in the form $a(x \pm b)^2 \pm c$.

(i) $2x^2 - 4x + 7$

(ii) $2x^2 + 12x + 11$

(iii) $3x^2 + 12x - 4$

(iv) $5x^2 - 40x + 72$

(v) $4x^2 + 24x - 16$

(vi) $9x^2 - 6x$

4 Write the following expressions in the form $a - (x + b)^2$, stating the values of a and b.

(i) $3 - 8x - x^2$

(ii) $1 - 2x - x^2$

5 Write the expression $7 + 8x - 2x^2$ in the form $a - b(x - c)^2$, stating the values of a, b and c.

6 Sketch these quadratic curves and state the co-ordinates of the vertex.

(i) $y = (x + 1)^2$

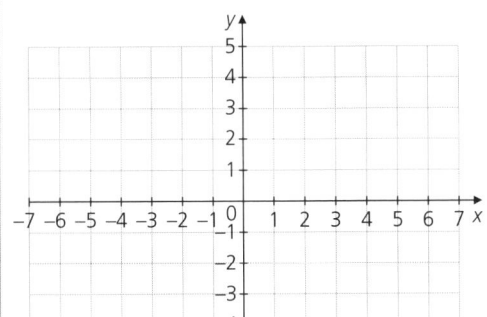

Vertex: (,   )

(ii) $y = (x + 4)^2 - 2$

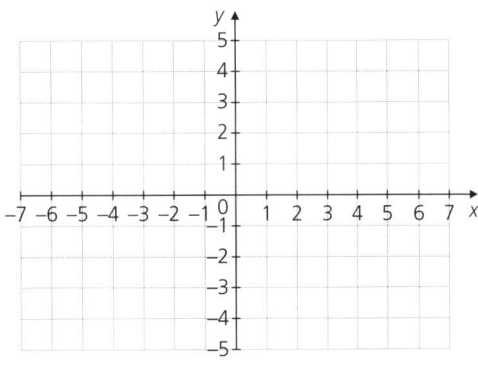

Vertex: (,   )

(iii) $y = -(x - 2)^2 + 1$

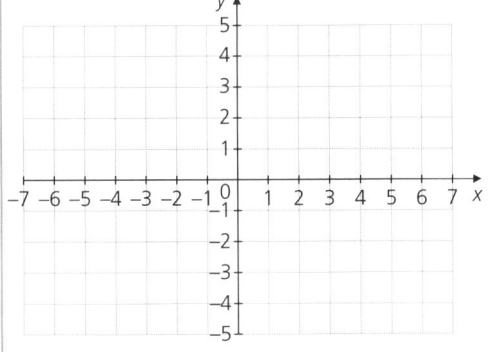

Vertex: (,   )

(iv) $y = 2(1 - 2x)^2 - 4$

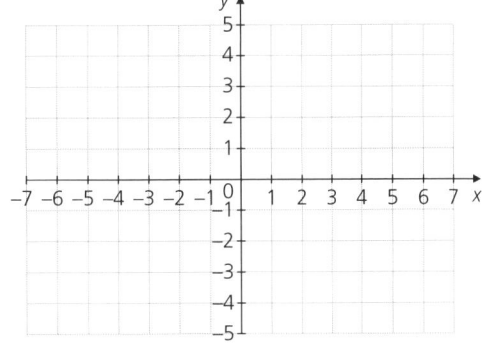

Vertex: (,   )

(v) $y = (2x + 1)^2 + 1$

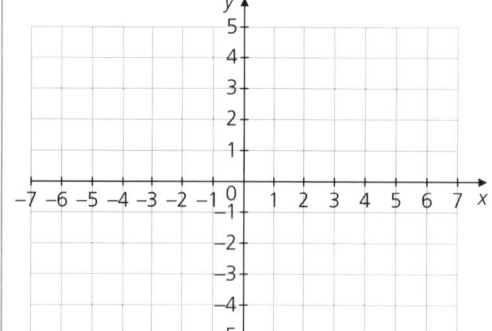

Vertex: (,   )

(vi) $y = (x + 1)(x + 3)$

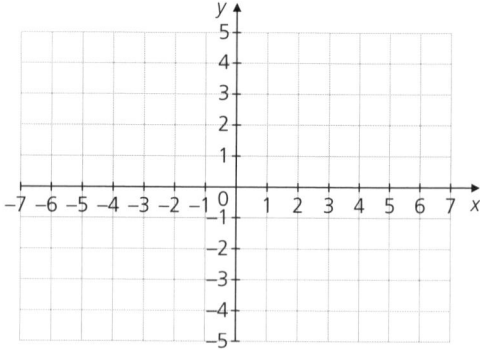

Vertex: (,   )

(vii) $y = (2 - x)(x + 1)$

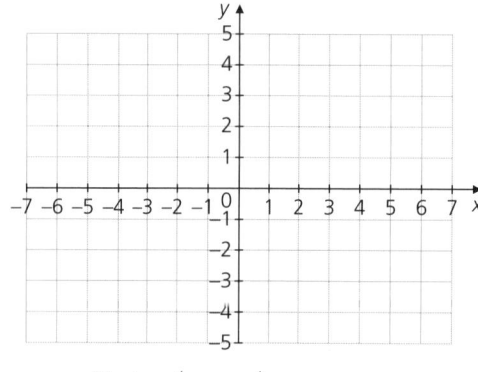

Vertex: (,   )

(viii) $y = -2(1 - x)^2 + 1$

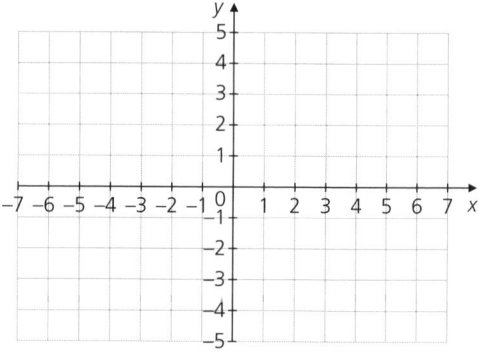

Vertex: (,   )

7 Write the equation of these graphs in the form $y = (x + b)^2 + c$.

(The coefficient of the x^2 term is 1.)

(i)    ....

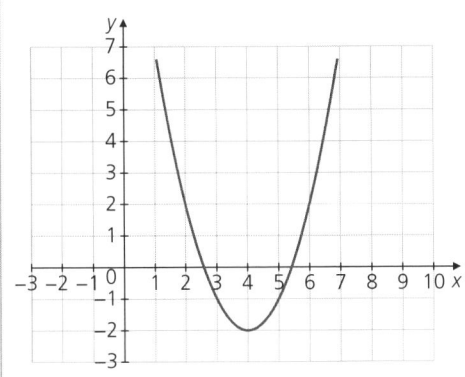

(ii)   ....

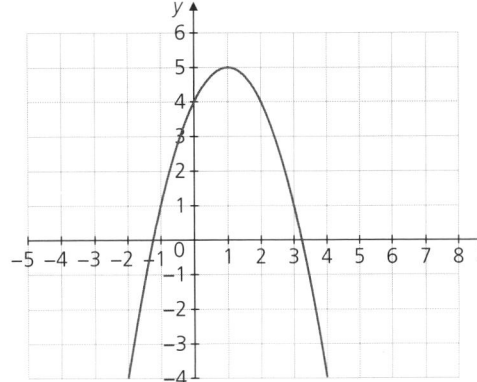

(iii)   ....

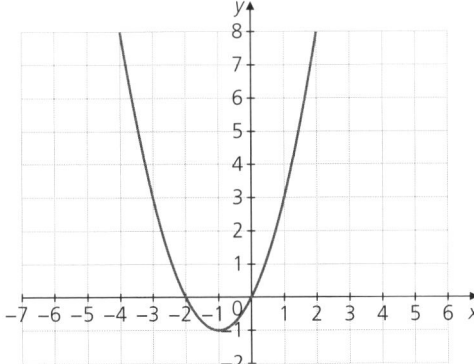

(iv)   ....

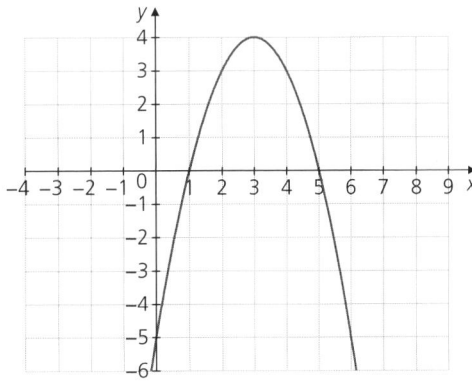

The quadratic formula

1 Solve the following equations using the quadratic formula.

(i) $x^2 + x - 5 = 0$

(ii) $x^2 - 5x = 7$

(iii) $1 - 3x^2 = 5x$

(iv) $12x = 6x^2 - 5$

2 Simplify these surds.

(i) $\sqrt{18}$

(ii) $\sqrt{75}$

(iii) $\sqrt{45}$

3 If $\sqrt{90} = a\sqrt{10}$ find the value of a.

4 Find the value of the discriminant for these quadratic equations, and hence state the number of real solutions for each equation.

(i) $x^2 - 2x + 4 = 0$

(ii) $x^2 + 4 = 0$

(iii) $-x^2 - 3x - 2 = 0$

(iv) $4x^2 + 4x + 1 = 0$

(v) $2 - x + 5x^2 = 0$

(vi) $-4x^2 + 3x = 0$

5 Find the value(s) of k for which these equations have one real solution.

(i) $kx^2 + 4x - 1 = 0$

(ii) $4 + kx + x^2 = 0$

(iii) $x^2 + kx + k - 1 = 0$ **(iv)** $kx^2 = kx + 1$

6 Find the value(s) of k for which these equations have two real solutions.

(i) $x^2 - 2x + k = 0$ **(ii)** $3x^2 - kx + 3 = 0$

(iii) $kx^2 - kx + 1 = 0$ **(iv)** $3 - 2kx^2 = 6x$

7 Find the value(s) of k for which these equations have no real solutions.

(i) $2x^2 - 2kx + 1 = 0$ **(ii)** $k - x + 9x^2 = 0$

(iii) $kx^2 - 4x + 2k = 0$ **(iv)** $2x + kx^2 = 1$

8 The quadratic equation $x^2 + mx + n = 0$, where m and n are constants, has roots 6 and −2.

(i) Find the values of m and n.

(ii) Using these values of m and n, find the value of the constant p such that the equation $x^2 + mx + n = p$ has one repeated root.

Simultaneous equations

1 Solve these equations simultaneously. Both equations are linear equations.

(i) $x - y = 4$
$x + 2y = 1$

(ii) $2x + 3y = 11$
$3x + y = -1$

(iii) $4x - \dfrac{1}{2}y = 6$
$3x - 2y = -15$

(iv) $2x + 3y = 3$
$y = 9 - 2x$

Inequalities

EXERCISE 1.6

1 Solve these linear inequalities.

(i) $3(x+4) \leqslant -15$

(ii) $1 - \dfrac{3x}{4} > 7$

(iii) $-4x - 1 < -x + 5$

2 Solve these quadratic inequalities.

(i) $x(x-1) > 0$

(ii) $2x(1-x) \geqslant 0$

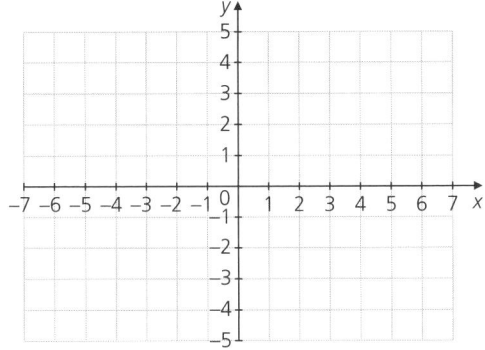

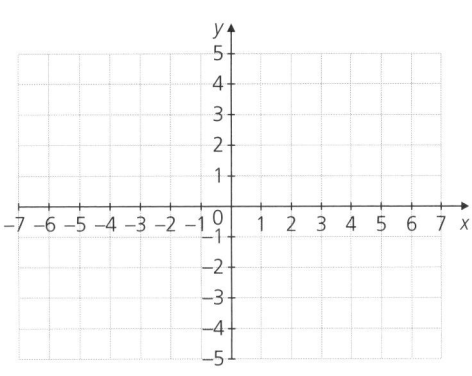

17

(iii) $(x + 1)(x - 1) > 0$

(iv) $x^2 < 9$

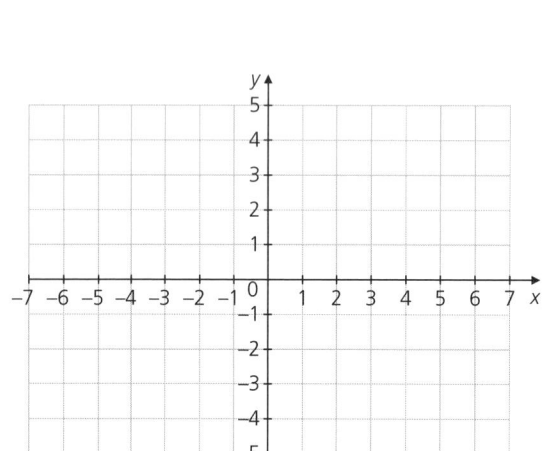

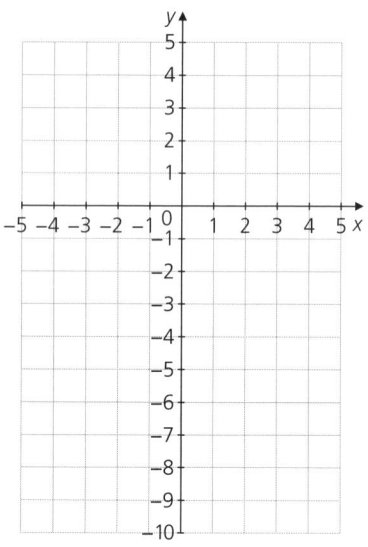

(v) $16 - x^2 \geqslant 0$

(vi) $x^2 + 5x > 0$

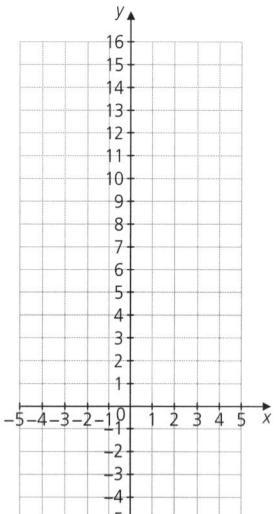

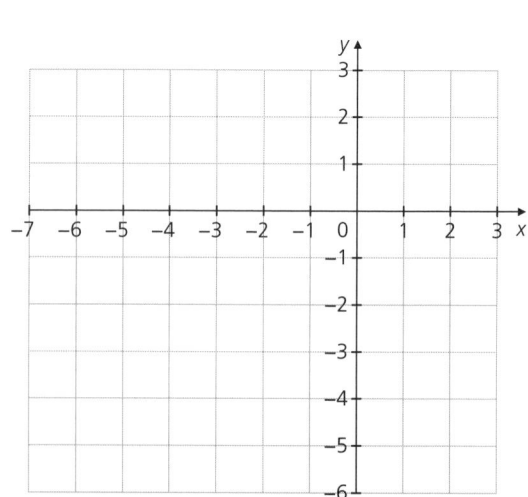

(vii) $x^2 - 2x - 8 \leqslant 0$

(viii) $x^2 + 5x - 14 > 0$

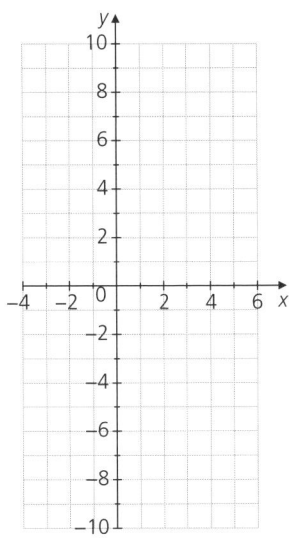

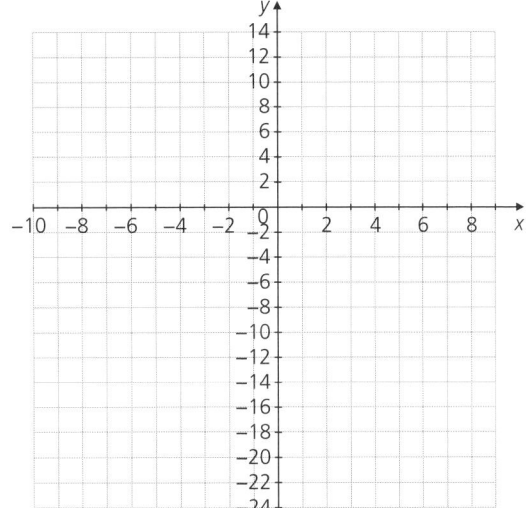

(ix) $(5 + x)(3 - 2x) < 0$

(x) $2x^2 - 5x - 3 < 0$

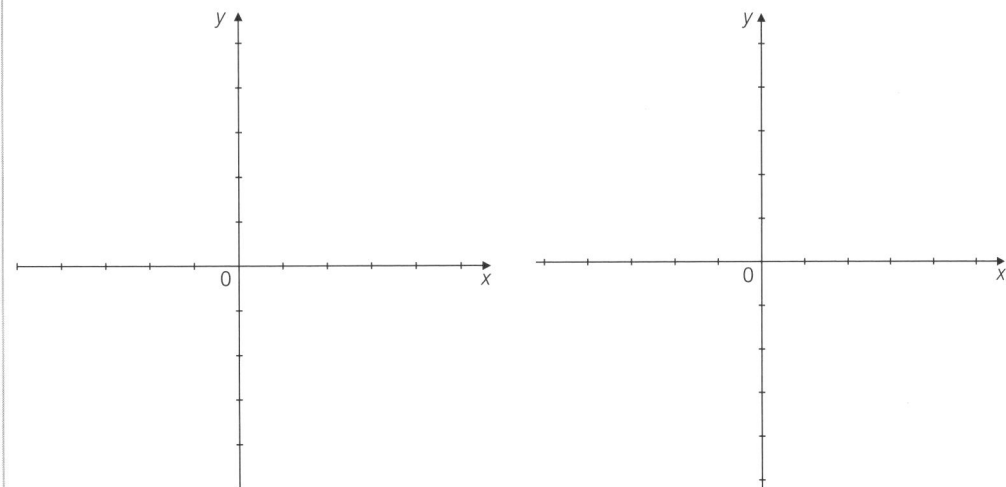

(xi) $3 - 8x - 3x^2 \geqslant 0$

(xii) $3x^2 > 6x$

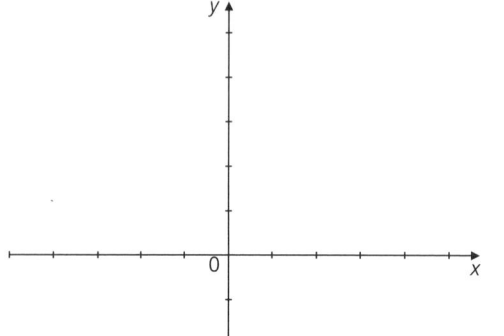

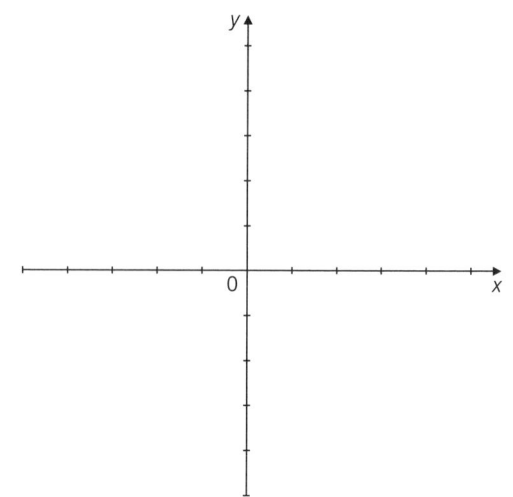

Stretch and challenge

1 Dan is shooting a basketball.

He stands at the point (0, 0) and releases the ball from the point (0, 8).

(i) Find the equation of the path of the ball above in the form $y = a - b(x + c)^2$.

(ii) Find another equation in the form $y = a - b(x + c)^2$ that the ball could follow to get in the hoop.

Stretch and challenge

2 (i) For a quadratic equation $ax^2 + bx + c = 0$ with roots α and β, show that

$$\alpha + \beta = -\frac{b}{a} \text{ and } \alpha\beta = \frac{c}{a}.$$

(ii) Using this fact, find the values of k for the equation $4x^2 + (k+2)x + 72 = 0$ such that one root is double the other.

(iii) The roots of the equation $3x^2 - 4x + 7 = 0$ are α and β.

Find the quadratic equation with roots $\dfrac{1}{\alpha}$ and $\dfrac{1}{\beta}$.

(iv) If α and β are the roots of the equation $x^2 - 2x + 3 = 0$, find the quadratic equation whose roots are α^3 and β^3.

3 Solve $9^x - 3^{x+1} - 54 = 0$.

4 Find all the possible values of k such that the equation $k2^x + 2^{-x} = 8$ has a single root.

Find the root in this case.

■ *Exam focus*

1 Write $2x^2 + 8x - 12$ in the form $a(x + b)^2 + c$, stating the values of a, b and c.　　　[3]

2 Express $4x - x^2$ in the form $a - (x + b)^2$, stating the numerical values of a and b.　　　[3]

3 Find the values of x such that $2x^2 + x - 1 \geqslant 0$.　　　[3]

4 (i) Write the expression $4x^2 + 32x + 70$ in the form $a(x + b)^2 + c$ and hence state the co-ordinates of the vertex of the graph of $y = 4x^2 + 32x + 70$.　　　[4]

(ii) Find the values of x when $y < 22$.　　　[3]

2 Co-ordinate geometry

The gradient of a line, The distance between two points, The mid-point of a line joining two points

EXERCISE 2.1

1 Find the gradient of the following straight lines.

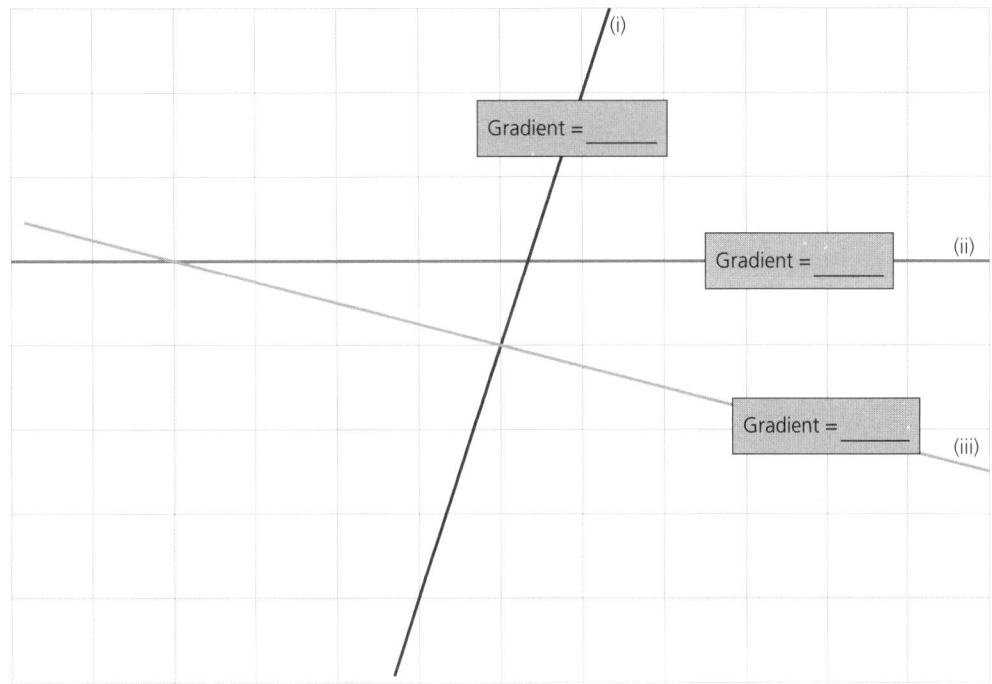

(i)

Gradient = _____

Gradient = _____ (ii)

Gradient = _____ (iii)

2 Given the co-ordinates of the end points of these lines, find the length, mid-point and gradient of each line.

(i) A(3, 1) and B(−1, −1)

(ii) C(12, −3) and D(8, 7)

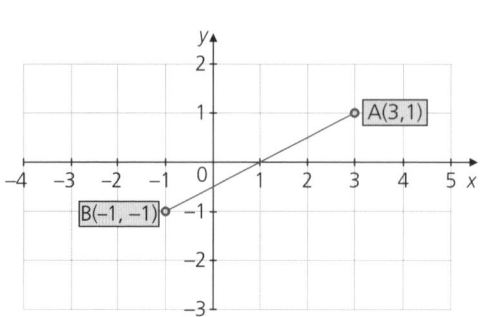

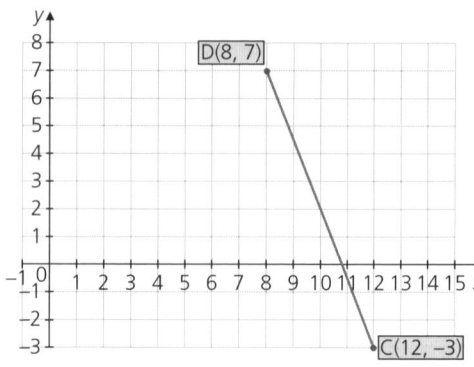

Length:

Length:

Mid-point:

Mid-point:

Gradient:

Gradient:

(iii) E(−5, 3) and F(3, −1)

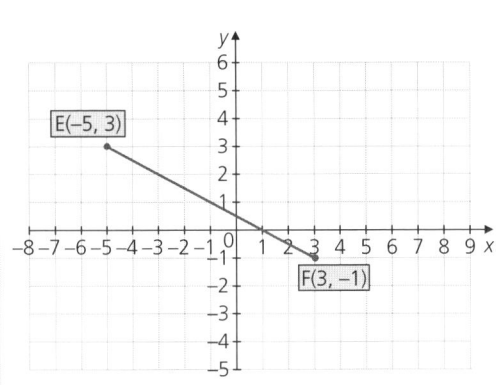

Length:

Mid-point:

Gradient:

(iv) G(−4, 3) and H(−10, −9)

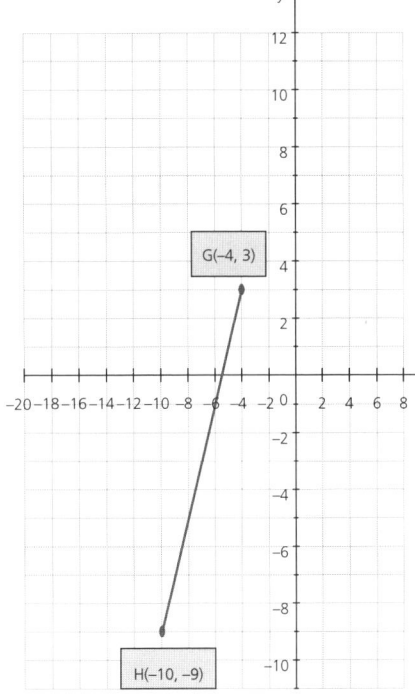

Length:

Mid-point:

Gradient:

3 The mid-point of A(−1, 5) and B(m, n) is (2, 5).

Find the value of m and n.

4 C is the point (−2, 1) and D is the point (x, 3).

Find the value of x if the gradient of the line CD is:

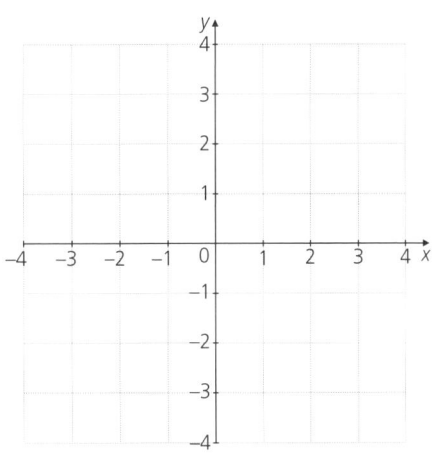

(i) 1

(ii) −1

(iii) $\frac{2}{3}$

5 (i) Find the angle these lines make with the positive x axis.

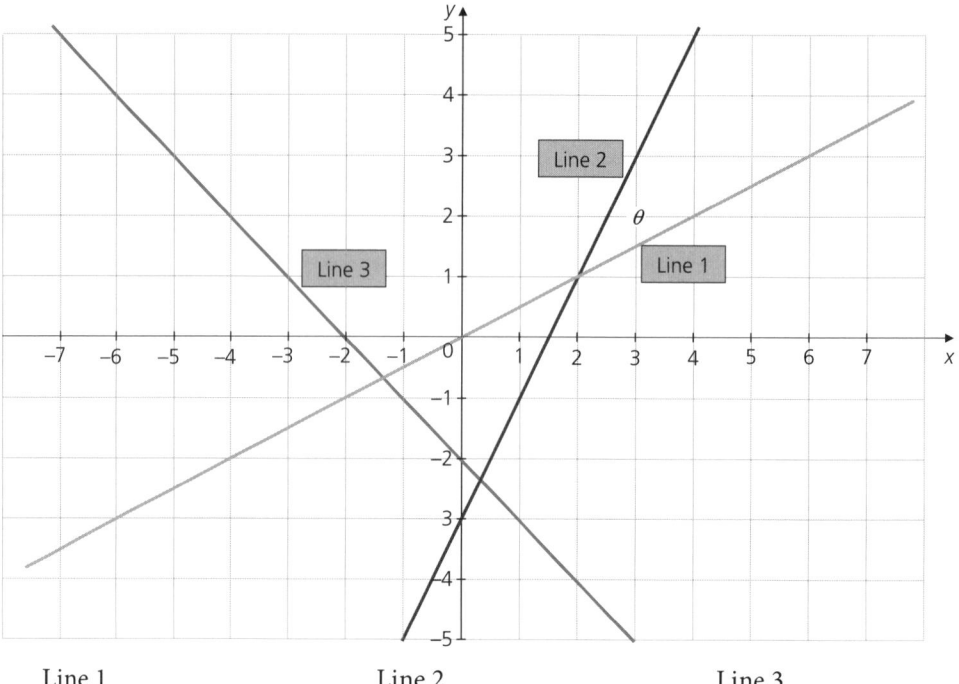

Line 1   Line 2   Line 3

(ii) Find θ, the angle between Line 1 and Line 2.

The equation of a straight line

1 Write down the gradient and y intercept of the following lines.

Sketch the lines on the axes below.

(i) $y = 3x - 1$

Gradient:

y intercept:

(ii) $y = -2x + 3$

Gradient:

y intercept:

(iii) $y = \dfrac{3x}{2} + 1$

Gradient:

y intercept:

(iv) $y = -\dfrac{x}{2} - 2$

Gradient:

y intercept:

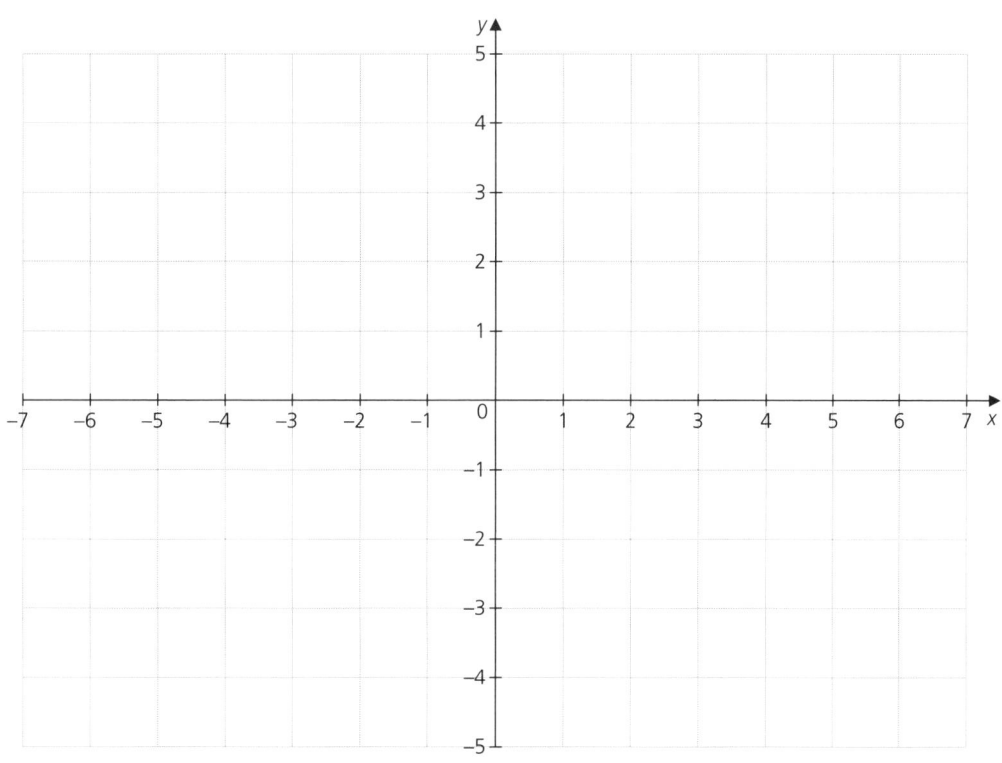

2 For each equation below:
- rearrange the equation so it is in the form $y = mx + c$
- hence state the gradient and y intercept
- draw the line on the axes below.

(i) $x + y = 2$ **(ii)** $3x - y = 2$

       Gradient: Gradient:

       y intercept: y intercept:

(iii) $2x + 4y - 9 = 0$ **(iv)** $3x - 2y = 8$

       Gradient: Gradient:

       y intercept: y intercept:

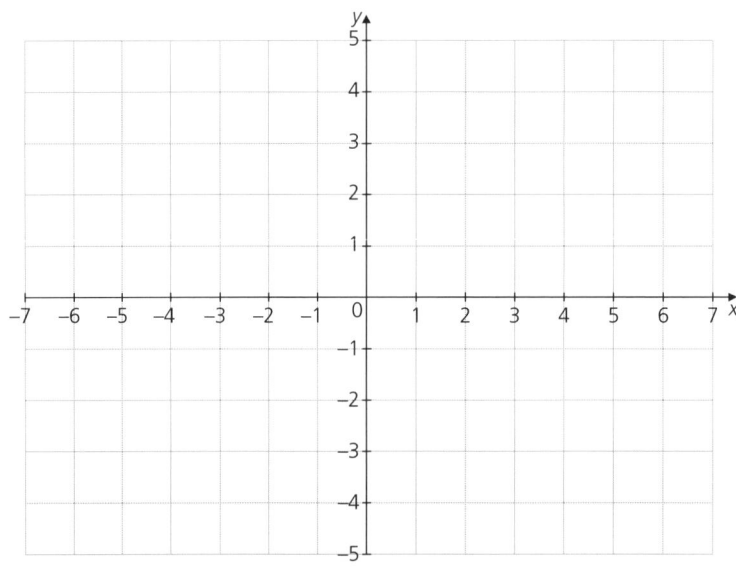

3 Rearrange the equations of these lines so they are in the form $ax + by + c = 0$.

(i) $y = -\dfrac{x}{3} - 2$ **(ii)** $y = \dfrac{4x}{5} + \dfrac{1}{3}$

Finding the equation of a line

1 Write down the equations of the lines below.

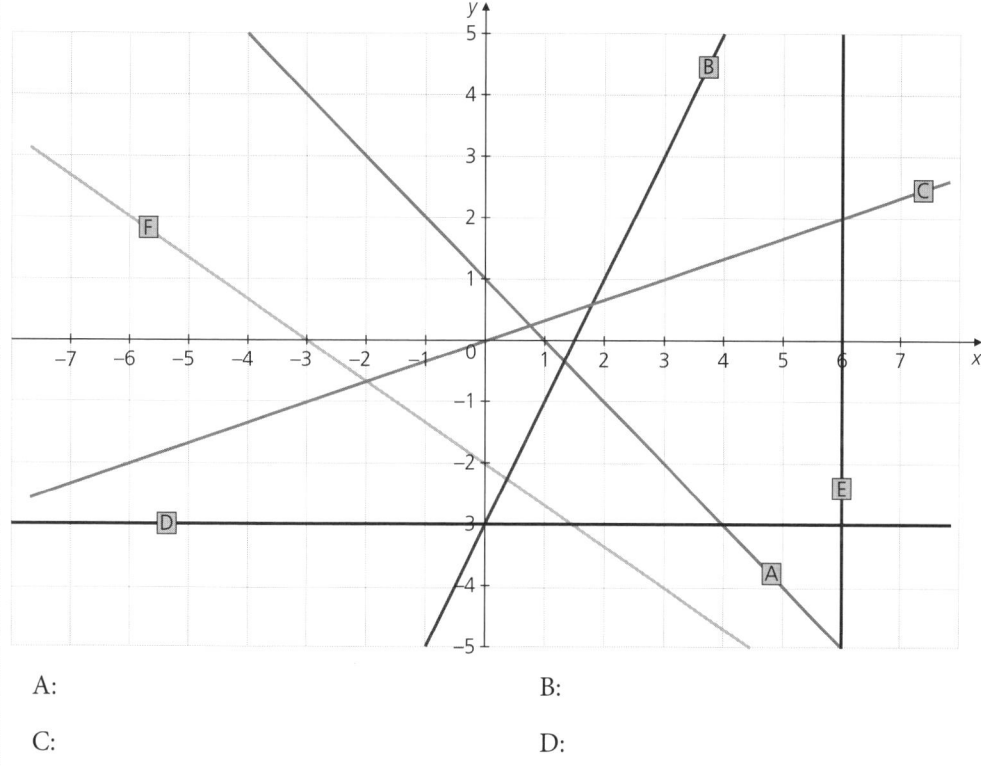

A: B:

C: D:

E: F:

2 Complete the table.

Gradient of line	Gradient of perpendicular
1	
	-4
$-2\frac{1}{3}$	
	$\frac{2}{3}$
0.3	

3 Find the equation of these lines.

(i) parallel to $2x - y = 1$ going through (4, 1)

(ii) perpendicular to $2x - y = 1$ going through (−3, 1)

(iii) Draw these two lines on the grid below.

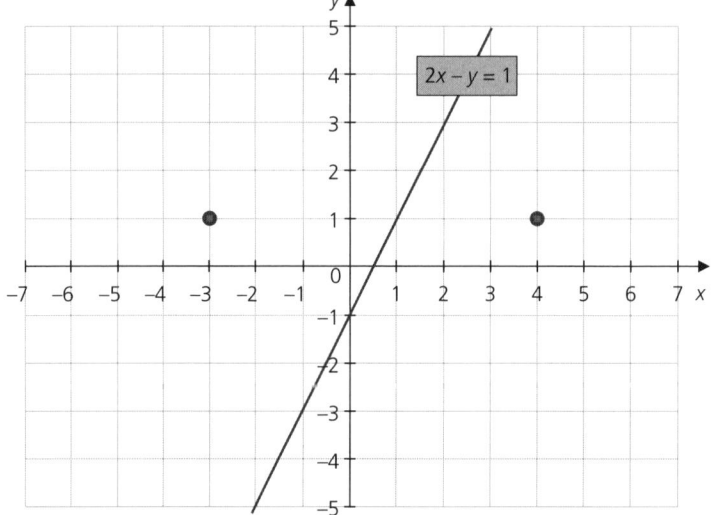

4 Find the equation of these lines.

Give your answers in the form $ax + by + c = 0$ where a, b and c are integers.

(i) parallel to $4x + 3y = 1$ going through (−2, 0)

(ii) perpendicular to $4x + 3y = 1$ going through (3, −1)

(iii) Draw these two lines on the grid below.

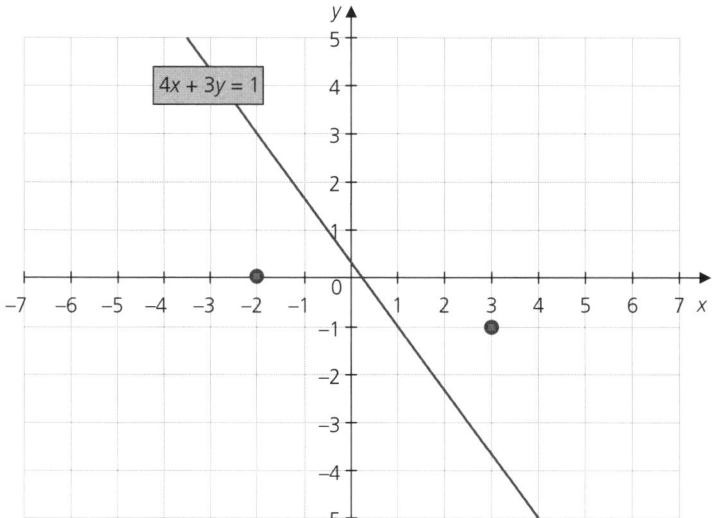

5 Points A, B, and C have co-ordinates (4, 1), (6, −2) and (−1, −9) respectively.

(i) Find the co-ordinates of the mid-point of AC.

(ii) Find the equation of the line through B perpendicular to AC.

Give your answer in the form $ax + by + c = 0$.

(iii) Draw the lines on the grid below.

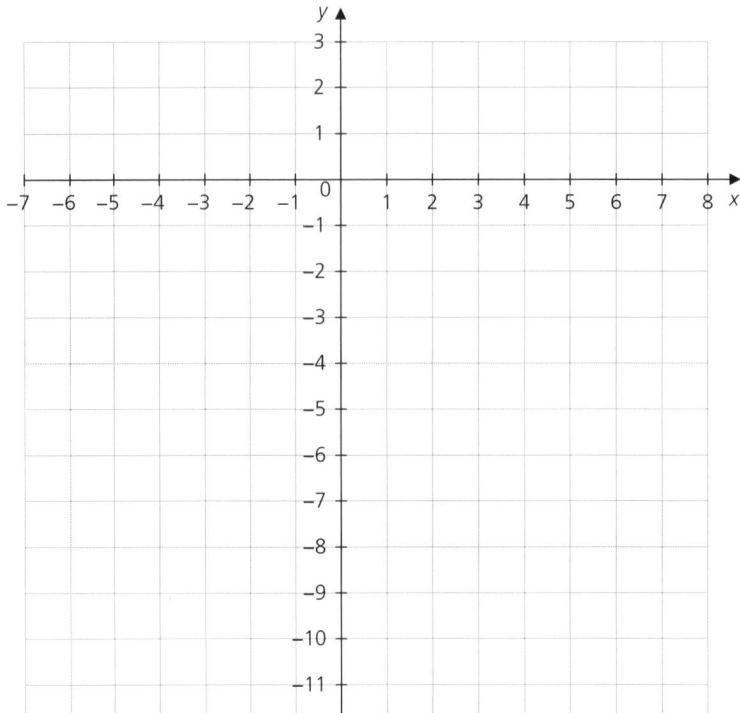

The intersection of two lines

EXERCISE 2.4

1 Find the area of the triangle formed by the straight line $3x + 2y = 8$ and the x and y axes.

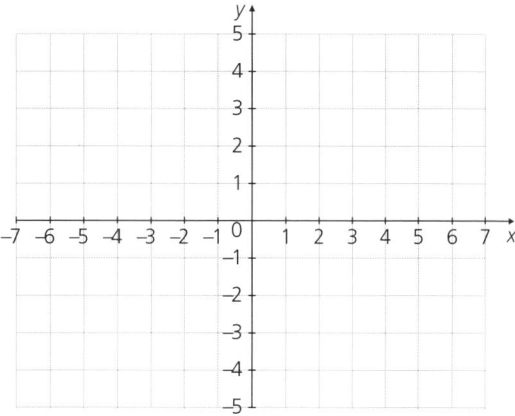

2 Find the equation of the perpendicular bisector of A(2, 4) and B(−6, 0)

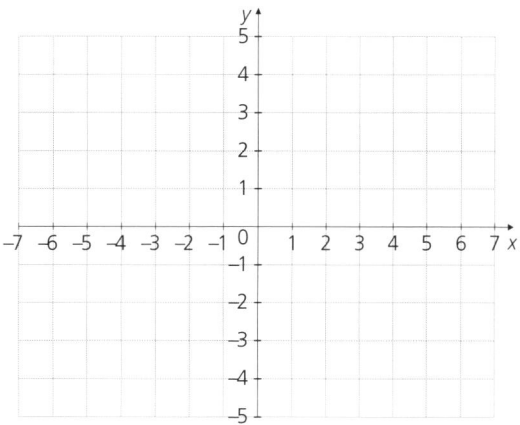

3 The line l_1 is given by $x − 3y = 6$. The point P is (−3, 2).

(i) Find the equation of the line perpendicular to l_1 that goes through P.

(ii) Find the perpendicular distance from the point P to l_1.

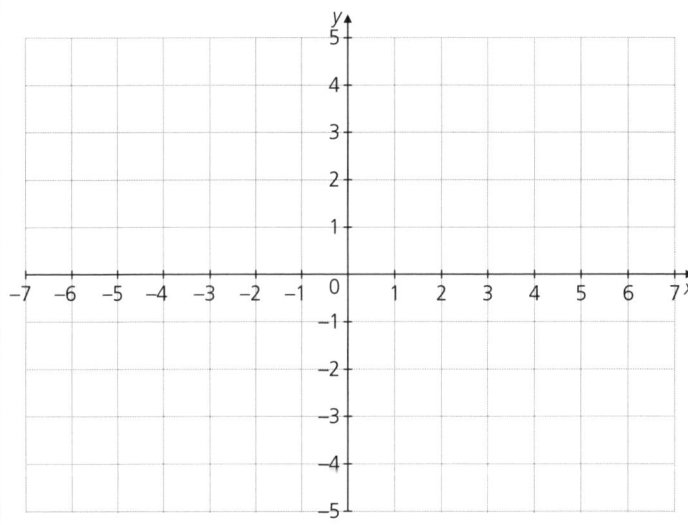

4 The median of a triangle is the line that joins a vertex to the mid-point of the opposite side.

A triangle is formed by the points A(−5, 2), B(2, 3) and C(4, −5).

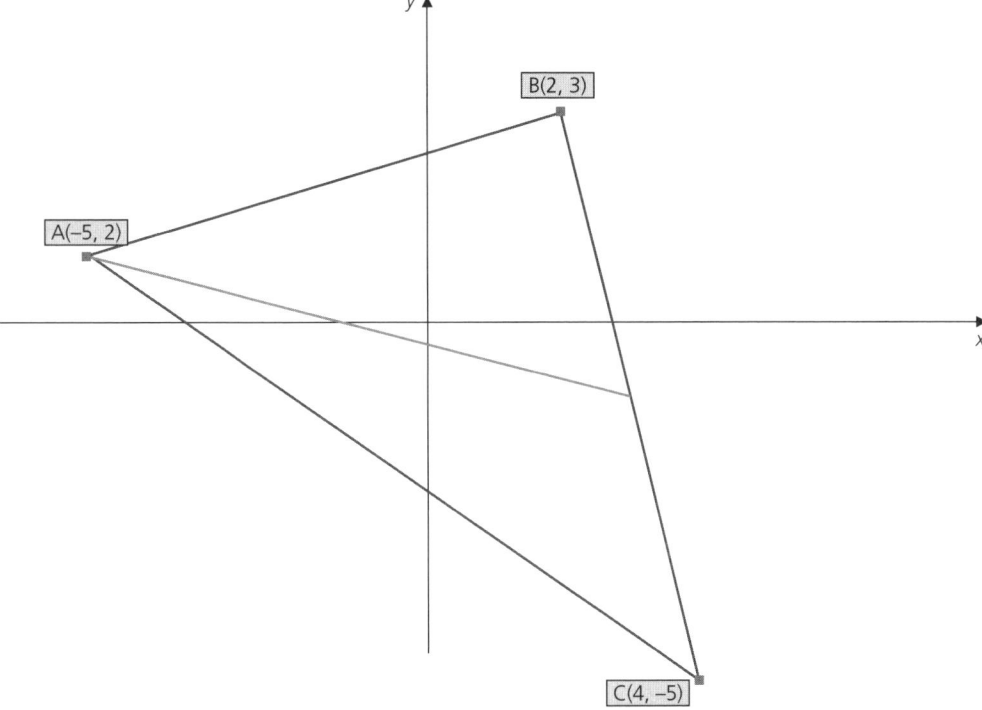

Find the equation of the median from the point A as shown.

5 Given that $ax - 3y + 1 = 0$ and $2x + by - 6 = 0$ are perpendicular lines, find the ratio $a:b$.

6 The diagram shows a rectangle with points A(−2, 0), B(6, 6), C(3, *b*) and D.

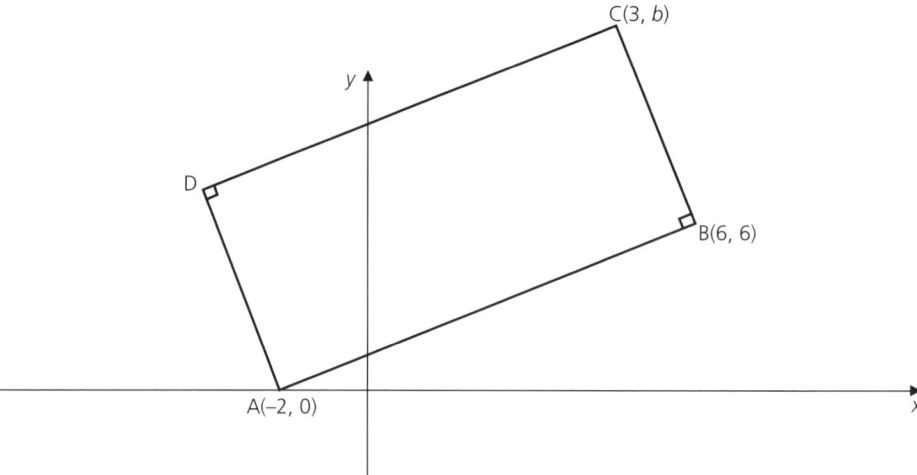

(i) Find the equation of BC and hence find the value of *b*, the *y* co-ordinate of the point C.

(ii) Find the equation of CD.

(iii) Find the equation of AD.

(iv) Find the co-ordinates of the point D.

The intersection of a line and a curve

EXERCISE 2.5

1 Find the co-ordinates of the point(s) of intersection of these lines and curves.

(i) $y = x^2 + 4x$
$y = x - 2$

(ii) $y = x^2 + 4x - 2$
$2x + y = -2$

(iii) $y = x^2 + 9$
$y - 4x = 5$

(iv) $xy = 3$
$2x + y = 7$

(v) $x^2 + 2y^2 = 9$
$x + y = 1$

(vi) $(x - 2)^2 + y^2 = 5$
$2x + y = 9$

2 Find the value of k such that the line $y - x = k$ is a tangent to the curve $y = x^2 + 3x + 9$.

3 The curve $y = 2 - x^2$ and the line $kx + y = 3$ intersect at two points.

Find the possible values of k.

4 Find the values of k such that the curve $xy = k$ has no points of intersection with the line $2x + y = 3$.

5 Find the value(s) of k that make the line $y - kx = -\sqrt{8}$ a tangent to the curve $x^2 - xy = k$.

6 Find the value(s) of the constant k for which the line $y + kx = 8$ is a tangent to the curve $y = 2x^2 + 3x + 10$.

7 A curve has equation $y = 2 + kx^2$ and a line has equation $y = kx + 1$, where k is a non-zero constant.

(i) Find the set of values of k for which the curve and the line have no common points.

(ii) State the value of k for which the line is a tangent to the curve and, for this case, find the co-ordinates of the point where the line touches the curve.

8 The line $y = 9 - 2x$ is a tangent to the curve $y = x(4 - x)$ at the point where $x = 3$, as shown.

Find the area of the triangle formed by the tangent, the normal and the x axis.

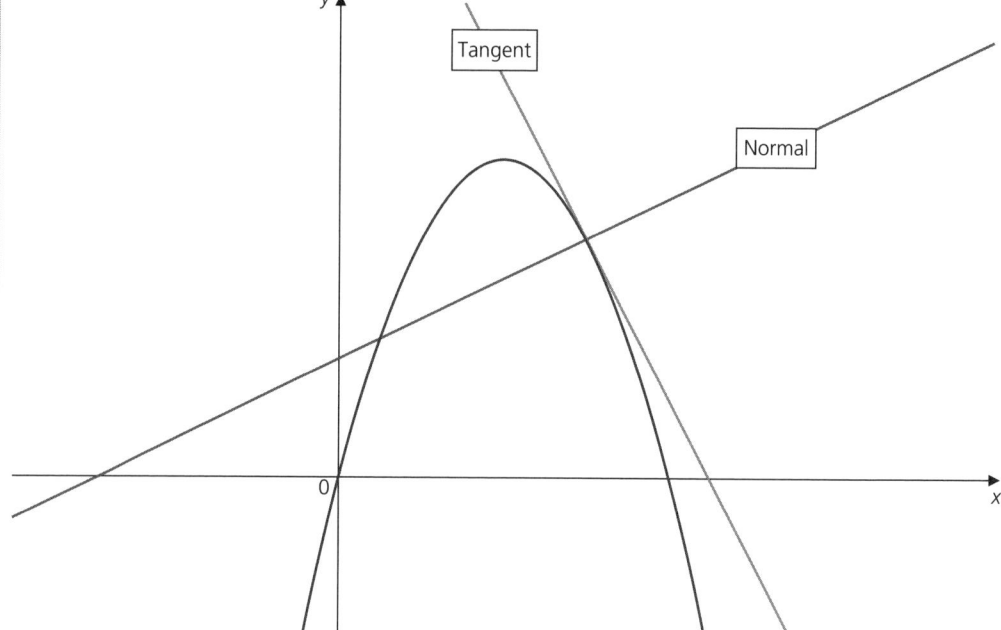

Stretch and challenge

1 The line $4x + y + 2 = 0$ is a tangent to the curve $y = 2x^2$ at the point A.

The normal to the curve at the same point also goes through the curve at the point B.

Find the length of AB.

2 The line $\dfrac{x}{a} - \dfrac{y}{b} = 4$, where a and b are positive constants, meets the x axis at M and the y axis at N.

Given that the gradient of the line is $\dfrac{1}{2}$ and that the length of MN is $\sqrt{720}$, find the values of a and b.

Stretch and challenge

3 The perpendicular bisector of A$(-6, 1)$ and B$(k, -3)$ passes through the y axis at -9.

Find the possible value(s) of the constant k.

4 (i) For what values of k does the line $y = kx + 1$ intersect the curve
$y = x^2 - 4x + 3$ on the x axis?

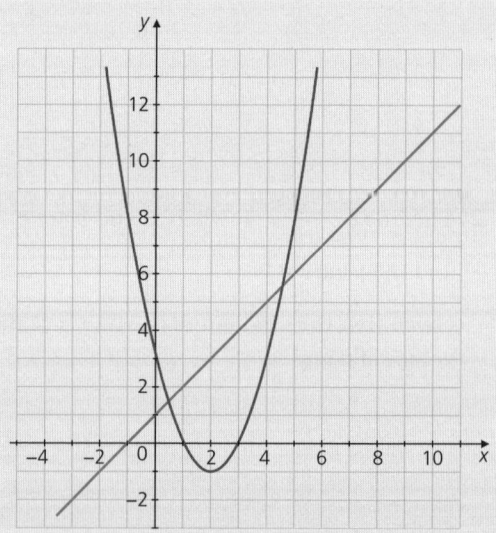

(ii) Find an expression for k in terms of a and b where the line $y = kx + 1$ intersects the curve $y = (x - a)(x - b)$ on the x axis.

(iii) Find an expression for k in terms of a, b and c where the line $y = kx + c$ intersects the curve $y = (x - a)(x - b)$ on the x axis.

5 For what values of k does the line $kx - 2y = -1$ intersect the curve $y = 2x^2 + x + 1$ at two points?

■ *Exam focus*

1 The points A(2, 5), B(−2, 2) and C(4, −1) are the vertices of a triangle.

The perpendicular from A meets the base BC at the point D.

Find the co-ordinates of the point D. [5]

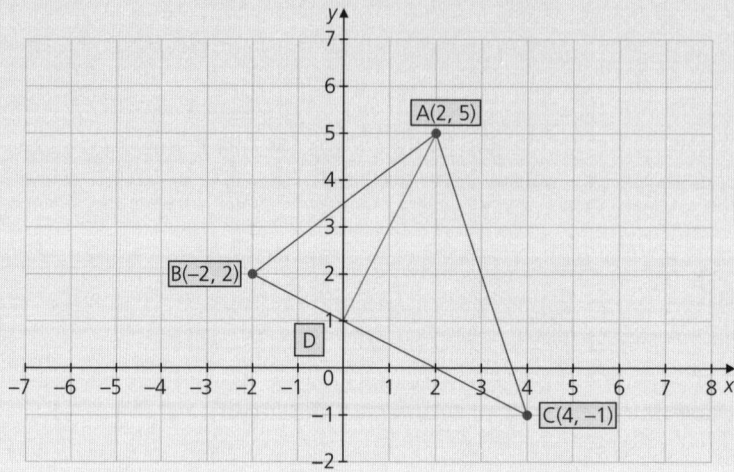

2 The diagram shows the points A(0, 6), B(8, 2), C(6, 8) and D.

The line BD is perpendicular to AB, and CD is parallel to AB.

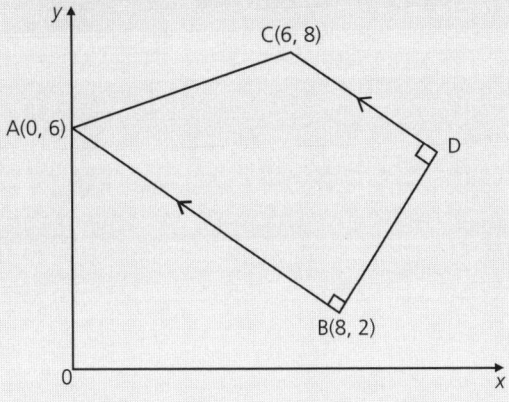

(i) Find the equation of the line CD. [2]

(ii) Find the equation of the line BD. [3]

(iii) Hence find the co-ordinates of the point D. [3]

3 Find the value of k such that the line $y - 2x = k$ is a tangent to the curve $y = x^2 - 2x - 1$.   [4]

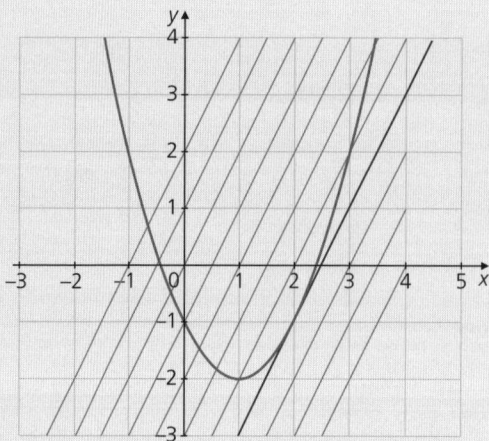

3 Sequences and series

Arithmetic progressions

EXERCISE 3.1

1 Find the term given in brackets for each of these arithmetic sequences.

   (i) 5, 11, 17, 23, ... (12th term)   **(ii)** −3, −8, −13, ... (15th term)

   (iii) $\frac{2}{3}, \frac{6}{5}, \frac{26}{15}, ...$ (7th term)   **(iv)** 0.85, 0.55, 0.25, ... (11th term)

2 Find the sum of these arithmetic progressions.

   (i) 21, 15, 9, ...   (9 terms)   **(ii)** −8, 0, 8, ...   (18 terms)

   (iii) 7, 9, 11, ... , 79   **(iv)** 0.05, 0.15, 0.25, ... , 4.05

3 In an arithmetic sequence, the fifth term is 34 and the tenth term is 43.

Find the first term and the sum of the first 20 terms of the sequence.

4 The first three terms of an arithmetic sequence are m, $2m + n$, $3m + 2n$, ...

Find the common difference and the tenth term in terms of m and n.

5 The third term of an arithmetic sequence is 12 and the sum of the first eight terms is 168.

Find the 14th term.

6 Paul buys a car on hire purchase and agrees to pay back the $3375 with weekly payments that are an arithmetic progression.

His first payment is $40 and the debt is paid in 30 weeks.

Find the 5th payment.

7 In an arithmetic progression, the first term is 25, the ninth term is 5 and the last term is −45.

Find the sum of all the terms in the progression.

8 The seventh term of an arithmetic progression is 32 and the sum of the first five terms is 130.

(i) Find the first term of the progression and the common difference.

(ii) The nth term of the progression is 56.

Find the value of n.

9 The first term of an arithmetic progression is four times the value of the fourth term.

The sixth term of the progression is four less than the fourth term.

Find the value of the eighth term.

Geometric progressions

1 Find the term given in brackets for each of these geometric sequences.

(i) $\frac{1}{2}$, 1, 2, 4, ...   (10th term)   (ii) $-16, 8, -4, 2, ...$ (8th term)

(iii) 0.1, 0.01, 0.001, ...   (9th term)   (iv) 6, 9, 13.5, ...   (12th term)

2 Find the sum of these geometric progressions.

(i)   3, 6, 12, 24, ...   (8 terms)   (ii) 81, 27, 9, 3, ...   (20 terms)

(iii) 2, −5, 12.5, −31.25, ...   (14 terms)   (iv) −240, 48, −9.6, 1.92, ... (10 terms)

3 Which of the sequences in question 2 have a sum to infinity?

For those sequences, calculate the sum to infinity.

4 Three terms of a geometric sequence are 32, x, 2. Find the value of x.

5 The third term of a geometric progression is 128 and the seventh term is 40.5.

Find the fifth term and the sum to infinity.

6 Find the fifth term of the geometric sequence with a first term of 8 and a sum to infinity of 12.

7 For what values of x does the sequence 3, $6x$, $12x^2$, ... have a sum to infinity?

8 The first three terms of a geometric sequence are $\sqrt{2}, 2, \sqrt{8}$.

Find the common ratio, r, and the sixth term of the sequence.

9 A car depreciates in value each year by 5%.

If the car was bought for \$45 000, find the value of the car eight years after it was bought.

10 The first, third and fifth terms of a geometric sequence are $\frac{5x+1}{2}$, $x + 2$ and $\frac{x}{2}$ respectively.

    (i) Find the value of x.

    (ii) Given that the common ratio of this geometric sequence is positive, calculate the sum to infinity of the series.

11 Jenny decides to start running every day.

    She has two plans to consider:

    Plan A: Run for 2 minutes the first day, and increase her running time by 30% every day.

    Plan B: Run for 1 minute the first day and increase her running time by t minutes every day.

    (i) Find the total time (in minutes) Jenny runs after 20 days if she chooses Plan A.

    (ii) Find the value of t such that the total time Jenny runs for the first 20 days under both plans is the same.

12 The first term of an arithmetic progression is 18 and the common difference is d, where $d \neq 0$.

The first term, the fourth term and the sixth term of this arithmetic progression are the first term, the second term and the third term, respectively, of a geometric progression with common ratio r.

(i) Write down two equations connecting d and r.

Hence show that $r = \frac{2}{3}$ and find the value of d.

(ii) Find the sum to infinity of the geometric progression.

(iii) Find the value of n such that the sum of the first n terms of the arithmetic progression is zero.

Binomial expansions

1 Write out the binomial expansions of the following expressions.

(i) $(x + 2)^4$

(ii) $(1 - 3x)^3$

2 Find the first three terms in the following expansions, fully simplifying each term.

(i) $(2x - 3)^8$

(ii) $\left(\dfrac{1}{x} + 2x\right)^6$

(iii) $(3a - b)^7$

(iv) $\left(\dfrac{1}{x^2} + x^3\right)^7$

3 Find the coefficient of the x^2 term in the expansion of $\left(2x + \dfrac{1}{x}\right)^{10}$.

4 Find the coefficient of the x^4 term in the expansion of $\left(\dfrac{x}{2} - \dfrac{3}{x}\right)^8$.

5 Find the term independent of x in the expansion of $\left(x^2 + \dfrac{5}{x^3}\right)^{15}$.

6 Find the coefficient of the x^2 term in the expansion of these expressions.

(i) $(2 - 3x)^6$

(ii) $(3 + 2x)(2 - 3x)^6$

7 Find the value of k such that the coefficient of the x^{-1} term in the expansion of $\left(kx + \dfrac{2}{x}\right)^5$ is 720.

8 Find the value of the constant b such that there is no term in x^3 in the expansion of $(1 + bx)(x + 2)^5$.

9 **(i)** Find the first three terms of $(3 + u)^6$ in ascending powers of u.

(ii) Use the substitution $u = x - x^2$ in your answer to part **(i)** to find the coefficient of the x^2 term in the expansion of $\left(3 + x - x^2\right)^6$.

10 In the expansion of $(2 - ax)^5$ the coefficient of the x^3 term is -1080.

Find the coefficient of the x^2 term.

11 (i) Find the first three terms in ascending powers of x for the expansion $(3 - 2x^2)^7$.

(ii) Find the coefficient of the x^4 term in the expansion of $(1 + x^2)(3 - 2x^2)^7$.

12 (i) Find the first three terms, in descending powers of x, in the expansion of $\left(x + \dfrac{3}{x}\right)^8$.

(ii) Find the coefficient of the x^6 term in the expansion of $(1 - x^2)\left(x + \dfrac{3}{x}\right)^8$.

13 The coefficient of the x^3 term in the expansion of $\left(k - \frac{1}{2}x\right)^6$ is 160.

Find the value of the constant k.

14 The coefficient of the x^2 term in the expansion of $(3 - x)^4 + (2 + ax)^5$, where a is a positive constant, is 554.

Find the value of a.

Stretch and challenge

1 Anna decides to reduce the time she spends on Facebook by the same number of minutes each week.

In week 7 of her plan she uses Facebook for 400 minutes.

After 30 weeks the total amount of time she used Facebook for was 1800 minutes.

For how many minutes did Anna use Facebook in week 1 of her plan?

2 The constant terms in the expansions of $\left(kx^3 + \dfrac{7}{x^3} \right)^6$ and $\left(kx^4 + \dfrac{m}{x^4} \right)^8$ are the same.

If k and m are positive constants, express k in terms of m.

Stretch and challenge

3 The sum of the first two terms of a geometric progression is −3.

The sum of the sixth and seventh terms is 729.

Find the common ratio and the first term of the progression.

4 In an arithmetic progression the 12th term is three times the value of the 6th term and the sum of the first 30 terms is 450.

Find the common difference and the first term.

5 A stamp collector buys two stamps.

The first is bought for $55 000 and its value depreciates by $2400 every year.

The second stamp depreciates by 4% every year.

After 10 years the stamps have the same value.

Find how much the second stamp was bought for.

6 (i) For what values of the positive constant a is the coefficient of x^3 in the expansion of $(x + a)^{10}$ the largest coefficient?

(ii) Which values of the positive constant a make the coefficient of x^{43} the largest coefficient in the expansion of $(x + a)^{100}$?

■ *Exam focus*

1 Find the coefficient of the x^3 term in the expansion of these expressions.

    (i) $(4 - 3x)^7$ [2]

    (ii) $(1 - x)(4 - 3x)^7$ [3]

2 Find the term independent of x in the expansion of $\left(\dfrac{2}{x} + x^2\right)^{12}$.   [3]

3 Find the value of the positive constant a such that the coefficient of the x^2 term in $\left(\dfrac{3}{x} + ax\right)^6$ is 2160.   [4]

4 An arithmetic sequence has a third term of 12 and a seventh term of 6.

Find the 21st term of the sequence. [3]

5 The fifth and tenth terms of an arithmetic sequence are −40 and −20 respectively.

Find the value of n such that the sum of the first n terms is zero. [4]

6 The second term in a geometric sequence is 9 and the fifth term is $1\frac{1}{8}$.

Find the sum to infinity of the sequence. [4]

4 Functions

The language of functions

1 Find the rule that links the x and f(x) values.

(i)

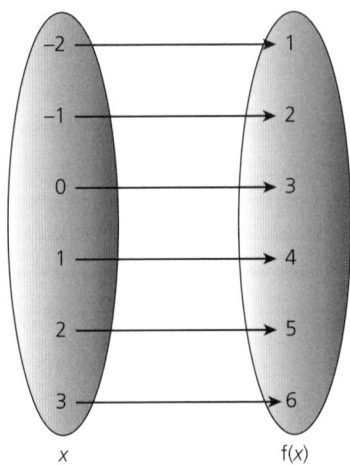

(ii)

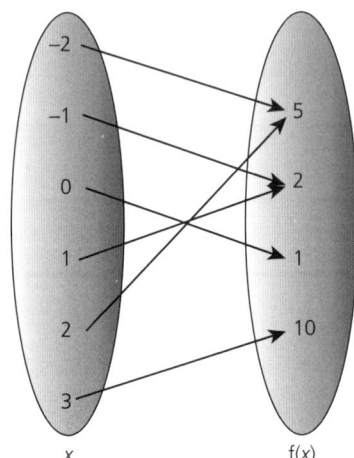

(iii)

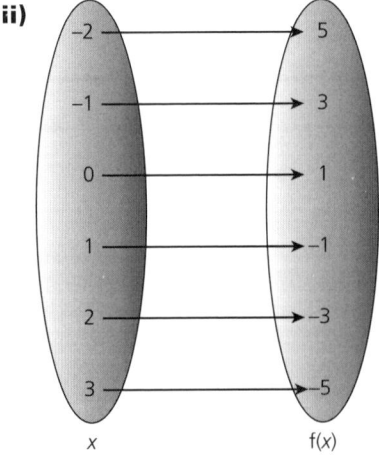

(iv)

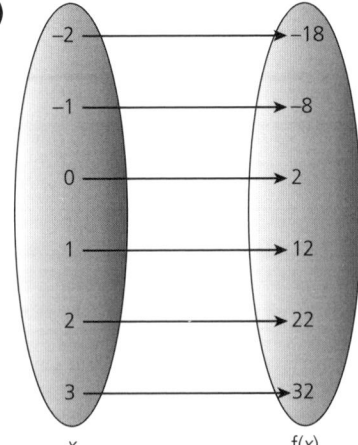

2 Which of the rules in question **1** are functions?

Are any of the rules one-to-one functions?

3 Are the following functions? If so, are they one-to-one functions?

(i)

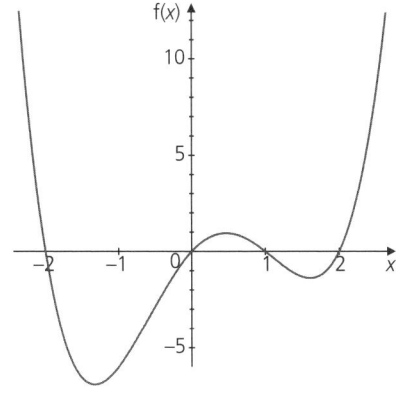

Function   Yes   No

One-to-one   Yes   No

(ii)

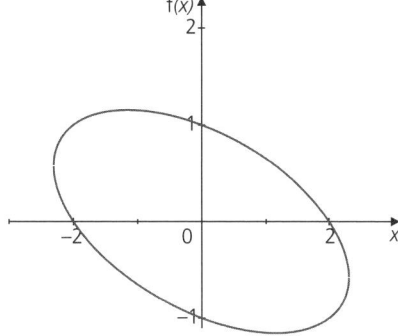

Function   Yes   No

One-to-one   Yes   No

(iii)

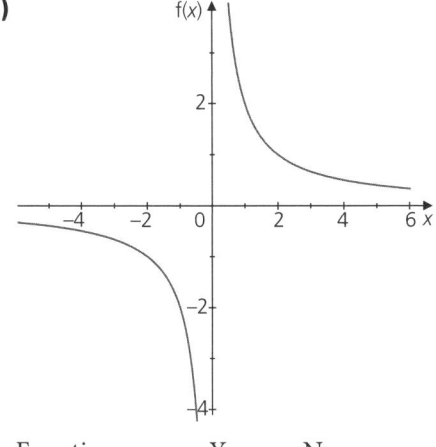

Function   Yes   No

One-to-one   Yes   No

(iv)

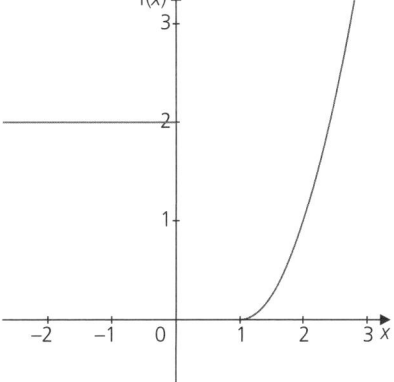

Function   Yes   No

One-to-one   Yes   No

4 For the function $f(x) = x^2 - 2x + 1$,

  (i) find f(−2)

  (ii) Find and simplify f(x + 1).

5 Are the following functions? If so, are they one-to-one functions?

(i)

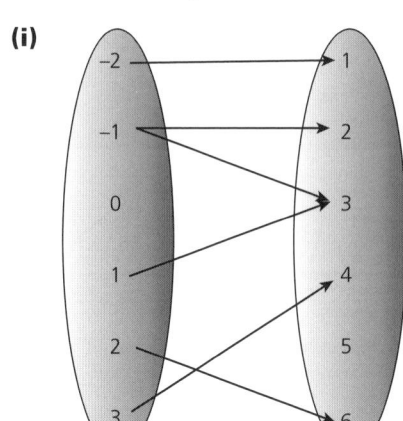

Function	Yes	No
One-to-one	Yes	No

(ii)

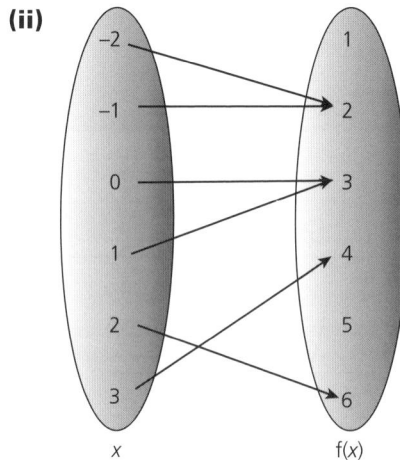

Function	Yes	No
One-to-one	Yes	No

(iii)

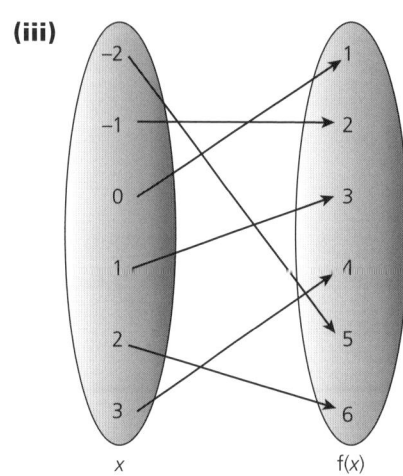

Function	Yes	No
One-to-one	Yes	No

(iv)

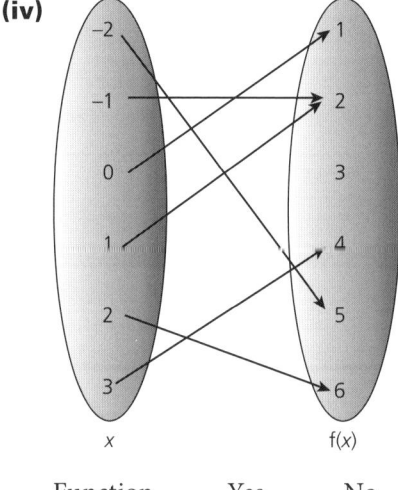

Function	Yes	No
One-to-one	Yes	No

6 Find the domain and range of these functions.

(i) $f(x) = (x - 2)^2 - 2$

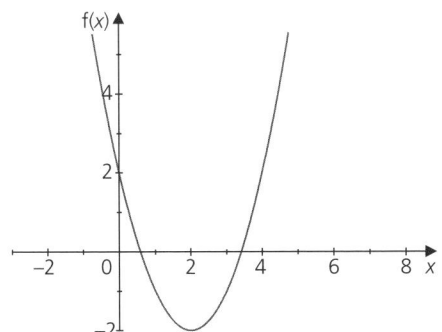

(ii) $f(x) = -(x + 1)^2 + 3$

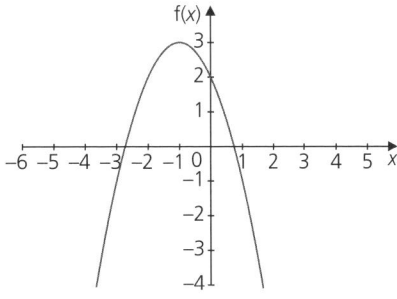

Domain:

Range:

Domain:

Range:

(iii) $f(x) = \sqrt{x - 1} + 2$

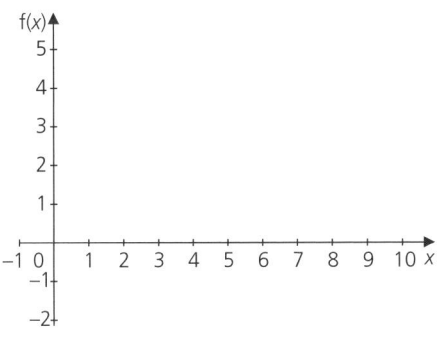

(iv) $f(x) = (x + 1)^3$

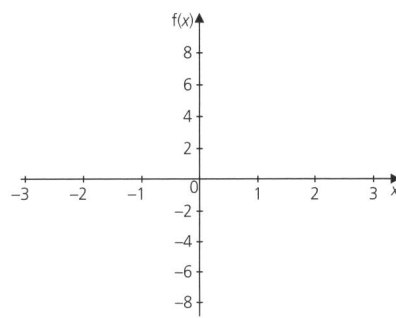

Domain:

Range:

Domain:

Range:

7 A function is defined by $f(x) = x^2 - 4x + 1,\ x \leqslant 1$.

Find the range.

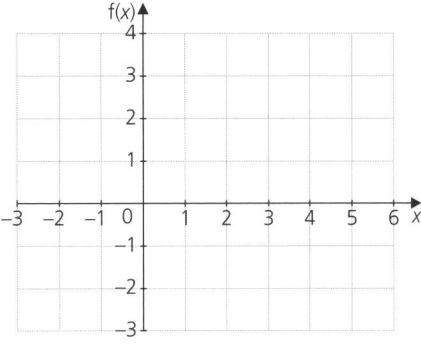

8 Find the domain and range of $g(x) = \sqrt{2 - \sqrt{x}}$.

9 Find the domain and range of $f(x) = 2x - x^2$.

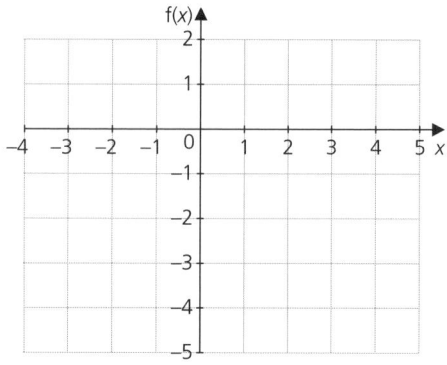

10 A function is defined by
$k(x) = 2x^2 + 8x + 3, x \geqslant -1$.

Find the range.

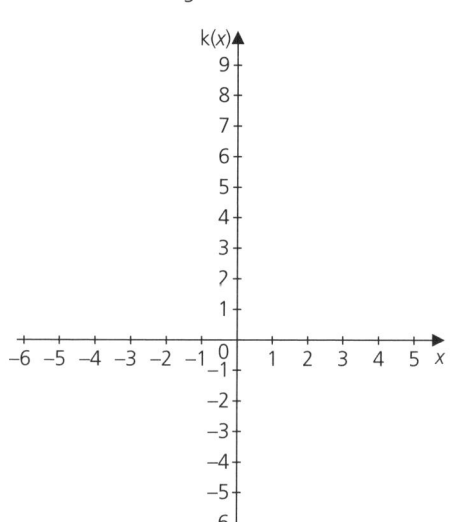

Composite functions

EXERCISE 4.2

1 For the functions $f(x) = 1 - x$ and $g(x) = 1 - x^2$, find the following.

(i) $f(-2)$

(ii) $g(-2)$

(iii) $fg(-2)$

(iv) $gf(-2)$

(v) $fg(x)$

(vi) $gf(x)$

(vii) $ff(x)$

(viii) $gg(x)$

2 The function f is defined as $f : x \mapsto ax + b$ for $x \in \mathbb{R}$ where a and b are constants.

It is given that $ff(x) = 9x - 4$.

Find the possible values of a and b.

3 The functions f and g are defined by

$$f : x \mapsto 3x + 2$$

$$g : x \mapsto 4 + \frac{2}{x}, \qquad x \neq 0$$

(i) Find and simplify the expression for $fg(x)$. **(ii)** Show that $gf(x) = \dfrac{12x + 10}{3x + 2}$.

(iii) Solve $ff(x) = -3$.

(iv) Find the value(s) of k such that the equation $g(x) = kx$ has two solutions.

4 Find the values of a and b for the function $g : x \mapsto b - ax$ given that $g(1) = -1$ and $gg(1) = 5$.

5 Functions f and g are defined by:

$$f : x \mapsto 3(x+2) \qquad \text{for } x \in \mathbb{R}$$

$$g : x \mapsto x^2 + 4x \qquad \text{for } x \geqslant -2$$

(i) Find the set of values of x which satisfy $fg(x) \geqslant g(x)$.

(ii) Find the set of values of x which satisfy $gf(x) \leqslant 45$.

Inverse functions

1 Find an expression for the inverse of these functions.

On the axes provided, sketch the graph of f(x) and f^{-1}(x), showing the co-ordinates of their point of intersection and the relationship between the graphs.

(i)   f(x) = 2x + 1   **(ii)**   f(x) = 5 − x

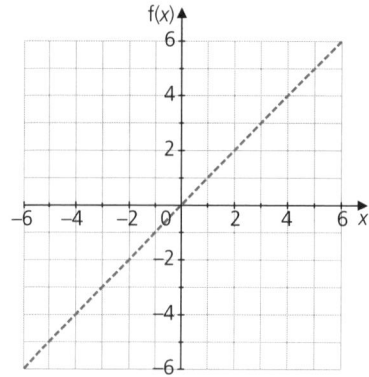 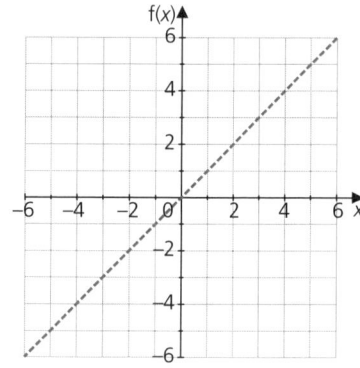

(iii)   f(x) = 1 − 5x **(iv)**   f(x) = $\dfrac{x}{2}$ − 1

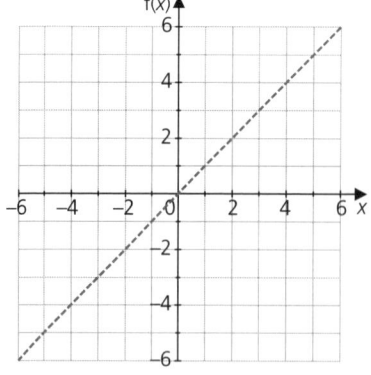 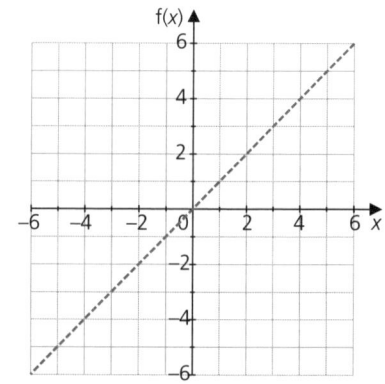

2 For each of the functions below:

- find the *least* value of k such that the domain $x \geqslant k$ means the function has an inverse
- sketch the graph and its inverse on the axes provided for the domain $x \geqslant k$
- find an expression for the inverse of the function
- state the domain and range of the function and its inverse.

(i) $f(x) = x^2 - 4$

(ii) $f(x) = x^2 - 4x$

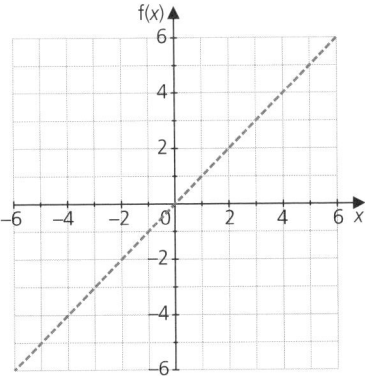

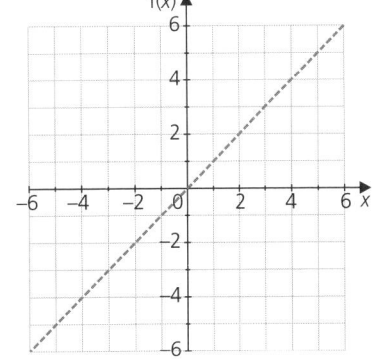

Domain f(x): $x \geqslant$

Range f(x):

Domain f^{-1}(x):

Range f^{-1}(x):

Domain f(x): $x \geqslant$

Range f(x):

Domain f^{-1}(x):

Range f^{-1}(x):

(iii) $f(x) = x^2 + 2x - 4$ **(iv)** $f(x) = 2x^2 + 8x + 7$

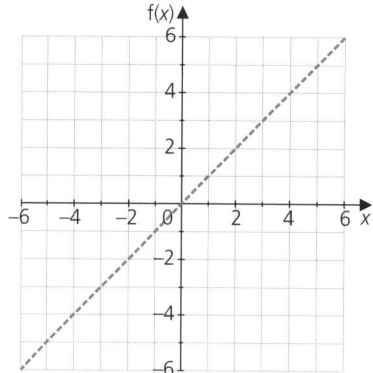

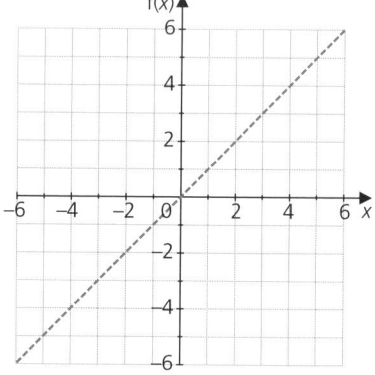

Domain f(x): $x \geqslant$ Domain f(x): $x \geqslant$

Range f(x):   Range f(x):

Domain f^{-1}(x):   Domain f^{-1}(x):

Range f^{-1}(x):   Range f^{-1}(x):

3 The function $g : x \mapsto x^2 + 3x + 1$ has an inverse when $x \leq k$.

Find the *largest* possible value of k.

Sketch the graph of $g(x)$ for $x \leq k$ and its inverse on the axes provided.

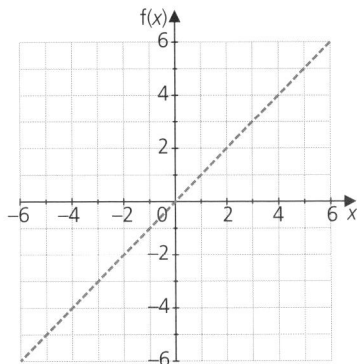

4 The function j is defined by $j : x \mapsto \dfrac{2}{3 - x}$ for $x \in \mathbb{R}, x \neq 3$.

Find an expression for $j^{-1}(x)$, the inverse of j.

5 Find an expression for the inverse of the function $f(x) = 2(x+1)^3 - 1$.

The diagram shows the graph of $f(x)$ and its inverse.

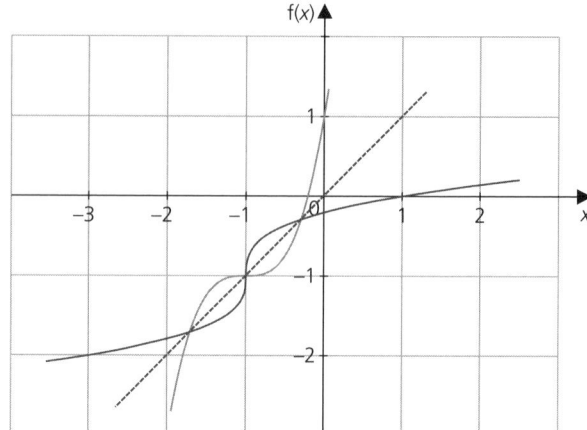

6 A function g is defined by $g(x) = \sqrt[4]{x^2 + 1} + 2$ for $x \geqslant 0$.

Find an expression for $g^{-1}(x)$ and write down the domain and range of $g(x)$ and $g^{-1}(x)$.

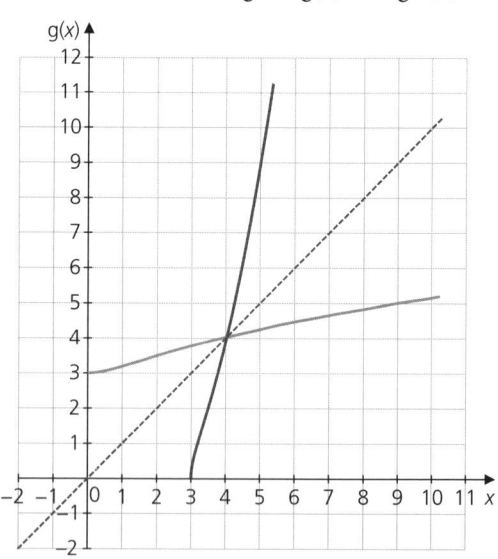

7 The graph of f(x) = cos x is shown for $0 \leqslant x \leqslant 2\pi$.

What is the maximum value of k such that $0 \leqslant x \leqslant k$ means that f(x) is one-to-one?

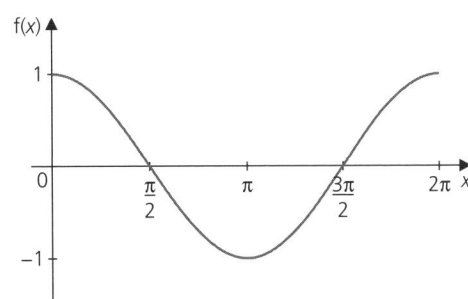

8 The diagram shows the function f defined for $0 \leqslant x \leqslant 14$ by:

$x \mapsto x^2$ for $0 \leqslant x \leqslant 2$

$x \mapsto \dfrac{1}{2}x + 3$ for $2 < x \leqslant 14$

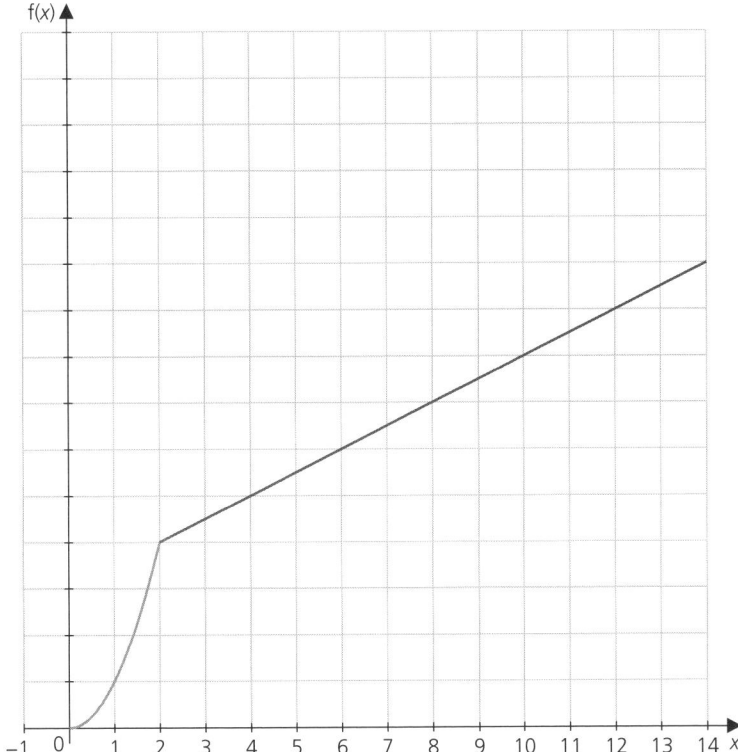

(i) State the range of f.

(ii) Sketch the graph of $y = f^{-1}(x)$ on the axes on the previous page.

(iii) Obtain expressions to define $f^{-1}(x)$, giving the set of values of x for which each expression is valid.

9 The function f is defined by $f : x \mapsto \dfrac{x+2}{4x-1}$, $x \in \mathbb{R}$, $x \neq \frac{1}{4}$.

(i) Show that $ff(x) = x$.

(ii) Hence, or otherwise, obtain an expression for $f^{-1}(x)$.

Stretch and challenge

1 (i) In Chapter 4 we looked at basic transformations of the curve $y = f(x)$ where $f(x) = x^2$.

Summarise the effects of each of the constants a, b, c and d on the graph of $y = af(b(x + c)) + d$.

If you have access to a graphical calculator or graphing software, this will help!

Constant	Effect on graph
a	
b	
c	
d	

(ii) The graph of $f(x) = x^2 - 3x$ is translated three units to the right and one unit up.

Find the equation of the new curve.

2 Functions f, g and h are defined as follows:

$f(x) = 2x + 1$

$g(1) = f(0)$

$g(x) = fg(x - 1), \qquad x \geqslant 1$

$h(0) = g(2)$

$h(x) = gh(x - 1), \qquad x \geqslant 1$

Find h(2).

Stretch and challenge

3 A function f is defined by $f(x) = 3^x$.

If $f(x + 1) + f(x - 1) = af(x)$, where a is a positive constant, find the value of a.

4 For a function $f(n) = an + b$, where a and b are integers, $f(2n - 1)$, $f(2n) - 1$ and $2f(n) - 1$ are three consecutive integers in some order.

Find all the possible functions $f(n)$.

■ Exam focus

1 A function f is defined by $f : x \mapsto x^2 - 4x + 2$ for $x \geqslant 2$.

Find the domain and range of $f^{-1}(x)$. [3]

2 A function f is defined by $f : x \mapsto ax + b$ for $x \in \mathbb{R}$ where a and b are constants.

It is given that $f(-1) = 1$ and $f(2) = 7$.

(i) Find the values of a and b. [3]

(ii) Solve $ff(x) = 1$. [3]

3 A function g is defined by $g : x \mapsto x^2 + 6x + 2, \ x \geqslant k$.

(i) Find the least value of k for which $g(x)$ has an inverse. [2]

(ii) Find the equation of the inverse in this case. [3]

4 Functions f and g are defined by:

$$f : x \mapsto 2 - 3x, \qquad x \in \mathbb{R}, x \geq 0$$

$$g : x \mapsto \frac{x+2}{2x+7}, \qquad x \in \mathbb{R}, x \neq -\frac{7}{2}$$

(i) Solve the equation $gf(x) = x$. [3]

(ii) Express $f^{-1}(x)$ and $g^{-1}(x)$ in terms of x. [3]

(iii) Show that the equation $g^{-1}(x) = x - 4$ has no solutions. [4]

(iv) Find and simplify an expression for $f^{-1}g(x)$. [2]

(v) Sketch in a single diagram the graphs of $y = f(x)$ and $y = f^{-1}(x)$, making clear the relationship between the graphs. [2]

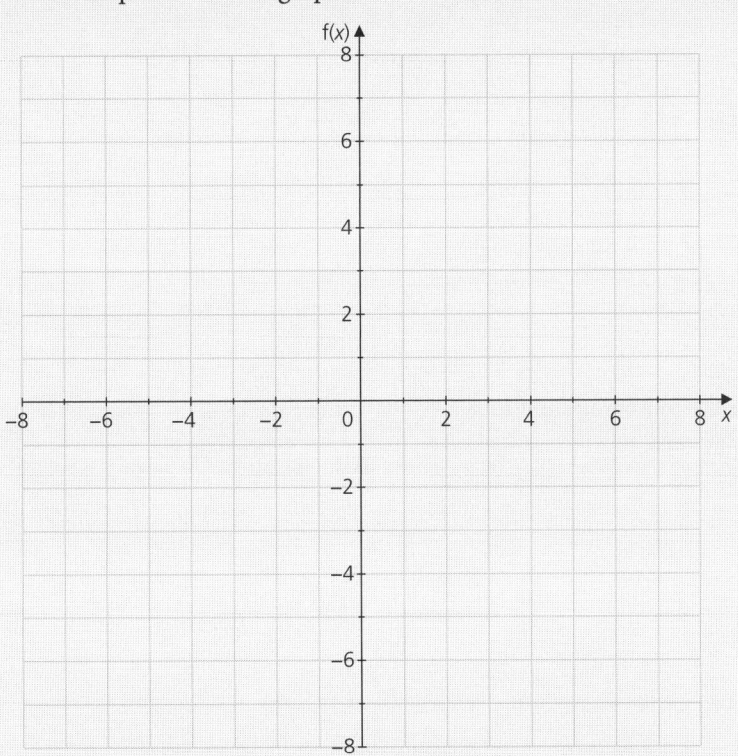

5 Differentiation

Differentiation by using standard results, Using differentiation, The second derivative

1 Differentiate the following functions.

(i) $y = 10x^2 - 3x + 1$

(ii) $y = 4 + x - 5x^4$

(iii) $y = \dfrac{5}{x^2}$

(iv) $y = \dfrac{2x^3}{3}$

(v) $y = \dfrac{4}{x} - \dfrac{x}{4}$

(vi) $y = \dfrac{1}{4x^3}$

(vii) $y = 2\sqrt{x} - 3x$

(viii) $y = \sqrt[3]{x} + \dfrac{2}{3x}$

2 Find $f'(x)$ for the following functions.

(i) $f(x) = \dfrac{2x + 1}{x^2}$

(ii) $f(x) = \dfrac{2x(x - 3)}{\sqrt{x}}$

(iii) $f(x) = \dfrac{\pi}{2}x + \dfrac{2}{\pi x}$

(iv) $f(x) = 6\sqrt[3]{x^5}$

3 Find $f'(9)$ if $f(x) = \dfrac{6}{\sqrt{x}} + 5x$.

4 Find the co-ordinates of the point on the curve $y = 3 - 8x - x^2$ where the gradient is 2.

5 The curve $f(x) = x^3 + x^2 + 2x - 1$ has two points where the slope of the curve is 3.

Find the co-ordinates of the points.

Tangents and normals

1 Find the equation of the tangent to the curve $y = 2x^4 + x^2 - 1$ at $(-1, 2)$.

2 Find the equation of the normal to the curve $g(x) = \dfrac{4}{\sqrt{x}} - 3$ at $x = 4$.

3 (i) The curve C has equation $y = \dfrac{1}{x^2} + x$.

The tangent to C at the point P(1, 2) intersects the x axis at the point Q and the y axis at the point R.

Find the length of PR.

(ii) The normal to C at the point P intersects the curve C again at the point S.

Find the co-ordinates of S.

4 (i) Show that the equation of the tangent to the curve $h(x) = 6x - x^2$ at the point P(2, 8) is $2x - y + 4 = 0$.

(ii) The normal to the curve at the point P meets the curve again at the point Q.

Find the co-ordinates of the point Q.

5 The gradient of the normal to the curve $y = x + \dfrac{k}{x}$ at the point where $x = -1$ is $\dfrac{1}{4}$.

Find the value of the constant k.

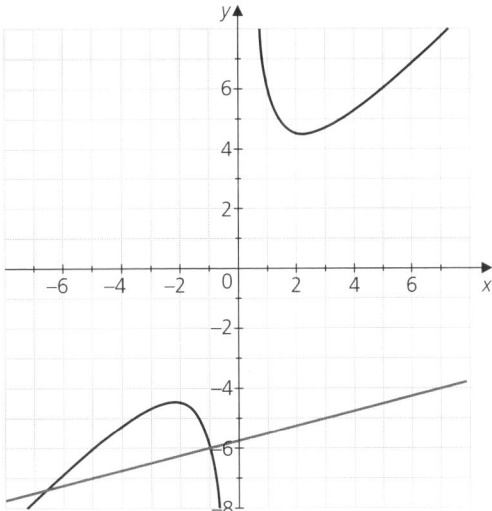

Maximum and minimum points, increasing and decreasing functions

EXERCISE 5.3

1 Find the co-ordinates of the stationary points and determine their nature for the following curves.

(i) $y = 2x^3 - 3x^2$

(ii) $f(x) = 4x + \dfrac{1}{x}$

(iii) $y = \sqrt{x} + \dfrac{1}{\sqrt{x}}$

2 Find the values of x where $g(x) = 3x^2 - 2x^3$ is increasing.

3 Show that the function $y = -x + x^2 - x^3$ is decreasing for all $x \in \mathbb{R}$.

4 Sketch the graphs of the following functions.

Use any information found in previous questions to help you.

(i) $y = 2x^3 - 3x^2$

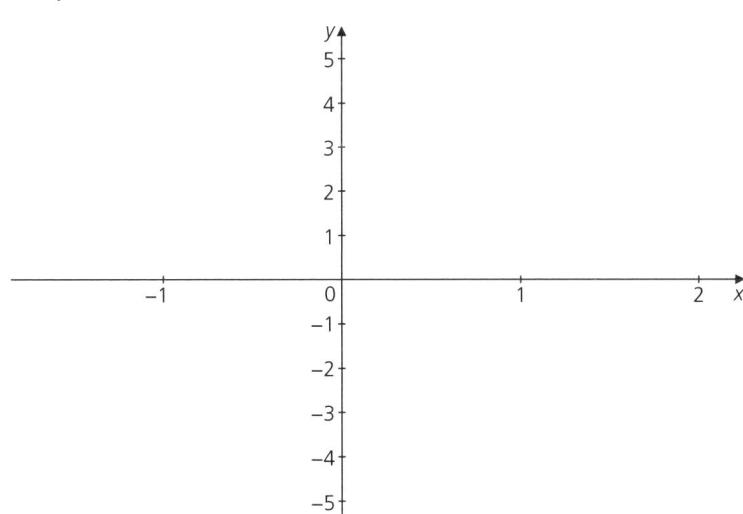

(ii) $g(x) = 3x^2 - 2x^3$

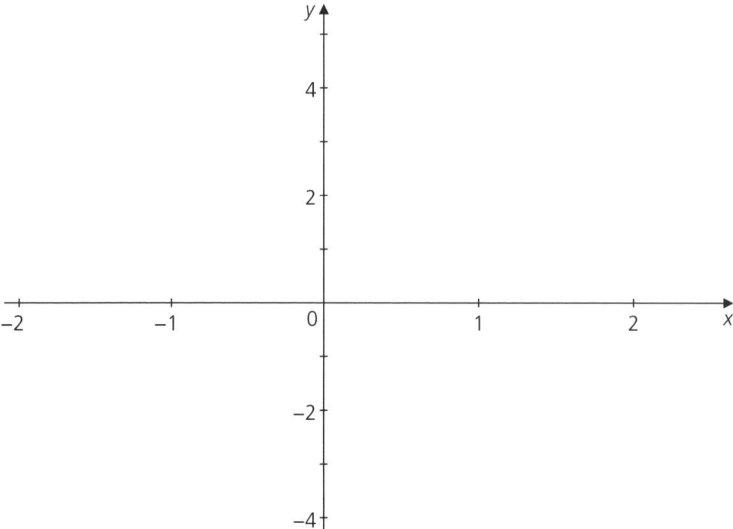

5 A curve $y = f(x)$ has a stationary point at the point $(1, -9)$.

It is given that $f'(x) = 3x^2 + kx - 8$, where k is a constant.

(i) Find the value of k.

(ii) Hence find the x co-ordinate of the other stationary point on the curve.

(iii) Find $f''(x)$ and determine the nature of each of the stationary points on $y = f(x)$.

Applications

EXERCISE 5.4

1 Two numbers have a sum of 18.

Find the maximum value of the product of the numbers.

2 A farmer wants to construct three pens for his chickens, as shown in the diagram.

Each pen has the same width, x.

The farmer has 36 m of material to make the pens.

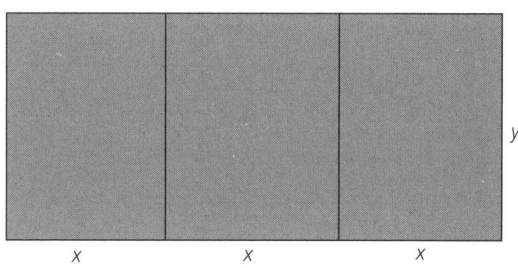

Find the dimensions of each pen so that the area of each pen is a maximum.

3 The diagram shows a glass window frame consisting of a rectangle of width x m and height y m and an equilateral triangle on top of the rectangle.

The perimeter of the window is 10 m.

(i) Show that the height, h, of the equilateral triangle is $\frac{\sqrt{3}}{2}x$.

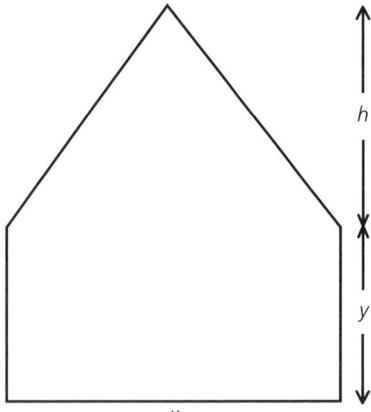

(ii) Show that the area of the window frame, A m², is given by $A = 5x + \frac{\sqrt{3}-6}{4}x^2$.

(iii) Given that x can vary, find the value of x for which A has a stationary value.

(iv) Determine whether this stationary value from **(iii)** is a maximum or a minimum.

4 A 5-litre paint can has radius r and height h. (5 litres = 5000 cm^3)

(i) Show that the surface area of the can, S, is given by

$$S = 2\pi r^2 + \frac{10000}{r}.$$

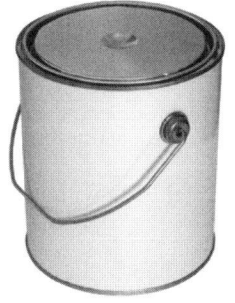

(ii) Given that r can vary, find the value of r for which S has a stationary value.

(iii) Determine whether this stationary value from **(ii)** is a maximum or a minimum.

5 An athletics track has a rectangular centre and a semi-circular area at each end, as shown.

The perimeter of the track is 400 m.

(i) Show that $b = 200 - \pi r$.

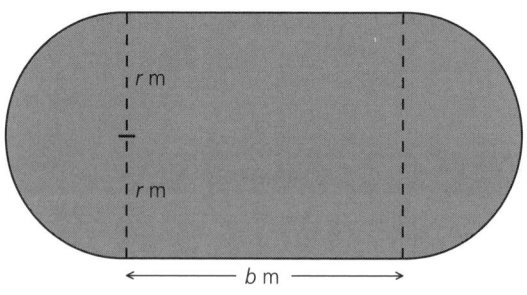

(ii) Find an expression for the area, A, of the rectangular section of the track, and hence find the values of r and b that maximise this area.

6 The post office will accept a package for shipment only if the sum of the girth (the distance around the middle) and the length is at most 120 cm.

For a package with a square end, what dimensions will give the package the largest possible volume?

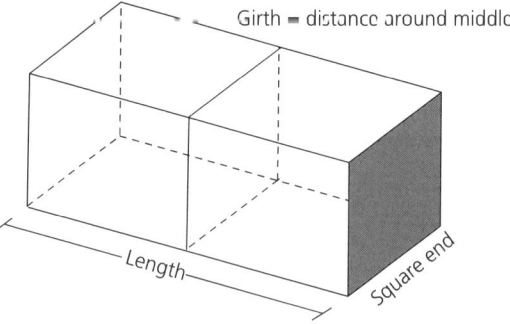

Girth = distance around middle

Length

Square end

7 The diagram shows a rectangle inscribed in a right-angled isosceles triangle with a hypotenuse of length 2.

What is the largest area the rectangle can have?

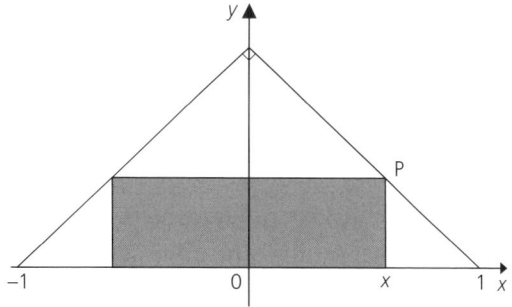

8 Find the volume of the largest right circular cone that can be inscribed in a sphere of radius 10 cm.

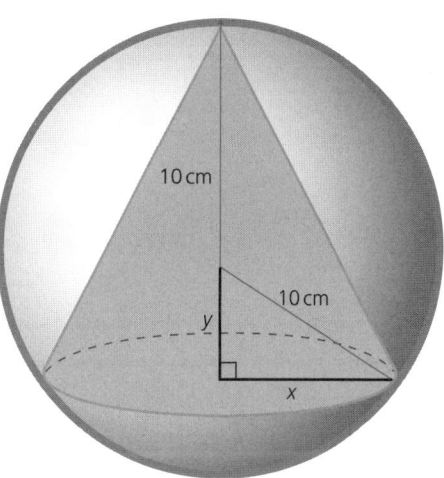

9 An open box is being constructed with a base length of 4 times the base width and a volume of $50\,\text{m}^3$.

If the materials used to build the box cost $12 per square metre for the bottom and $5 per square metre for the sides, what are the dimensions of the least expensive box?

The movie 'Stand and Deliver' is the true story of Maths teacher Jamie Escalante who challenges his class to step up from the basic maths course they are studying to sit the AP Calculus exam. Due to their unexpectedly excellent results and similar answers to one question, the students were required to re-take the exam.

This question is similar to the question that caused the re-take of the AP Calculus exam for Jamie Escalante's students.

The chain rule

EXERCISE 5.5

1 Use the chain rule to find $\dfrac{dy}{dx}$ for the following functions.

(i) $\quad y = (x - 4)^7$

(ii) $\quad y = (3x + 2)^8$

(iii) $y = (5 - 2x)^9$

(iv) $y = \sqrt{3 + 4x}$

(v) $\quad y = \sqrt[3]{1 - 9x}$

(vi) $y = \dfrac{2}{x + 3}$

(vii) $y = \dfrac{4}{(1 - 2x)^2}$

(viii) $y = \dfrac{5}{\sqrt{5x + 3}}$

2 Find the equation of the tangent to $f(x) = 3\left(1 - \dfrac{x}{6}\right)^4$ when $x = 12$.

3 Find the co-ordinates of the stationary point(s) of the function $y = \dfrac{8}{x^2 - 4x}$.

4 The function g is defined by $g : x \mapsto 3(x + 2)^3 - 5, x > -2$.

Obtain an expression for $g'(x)$ and use your answer to explain why g has an inverse.

5 (i) Find the value of the constant k such that the curve $y = \dfrac{k+1}{2x+3} + x$ has a gradient of 2 when $x = -1$.

(ii) When $k = 1$, find the values of x where the function is increasing.

6 A curve has the equation $y = \dfrac{2}{x-1} + 2x$.

(i) Find $\dfrac{dy}{dx}$ and $\dfrac{d^2y}{dx^2}$.

(ii) Find the co-ordinates of the maximum point A and the minimum point B on the curve.

7 The area of a square is increasing at a rate of 8 cm² per second.

Find the rate the length of the side is increasing at the instant when the side length is 20 cm.

8 Water is being pumped into a cylindrical tank of radius 1.2 m at a rate of 0.5 m³ per minute.

Find the rate at which the water level is rising.

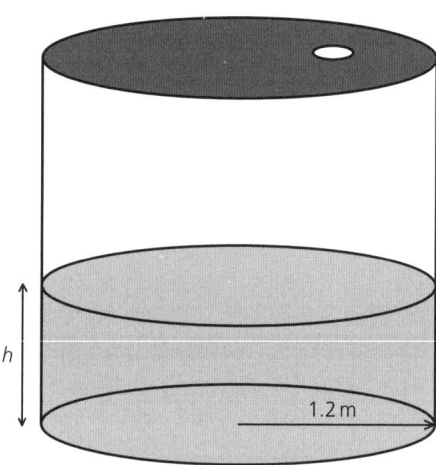

h

1.2 m

9 The equation of a curve is $y = (0.2x - 1)^{10}$.

A point with co-ordinates (x, y) moves along the curve in such a way that the rate of increase of x has the constant value 0.08 units per second.

Find the rate of increase of y at the instant when $x = 10$.

10 A spherical balloon is being blown up at a rate of $1500\,\text{cm}^3$ per second.

Volume of a sphere: $V = \frac{4}{3}\pi r^3$

Surface area of a sphere: $S = 4\pi r^2$

After 5 seconds find the rate of increase of:

(i) the radius, r

Hint:

$$\frac{\mathrm{d}r}{\mathrm{d}V} = \frac{1}{\dfrac{\mathrm{d}V}{\mathrm{d}r}}$$

(ii) the surface area, S.

11 A horse is training on a circular track of radius 200 m as shown in the diagram.

The horse starts at the point A and gallops to B at a constant 10 metres per second.

It is followed by a camera at the centre of the field at O.

Find the rate of change, in radians per second, of the angle θ during this time.

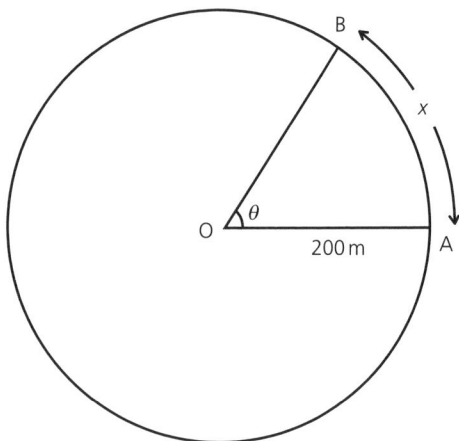

12 The diagram shows a water trough.

(i) When the depth of water in the trough is h cm, show that the trough contains a volume of $\frac{45}{2}h^2 + 3600h$ cm^3.

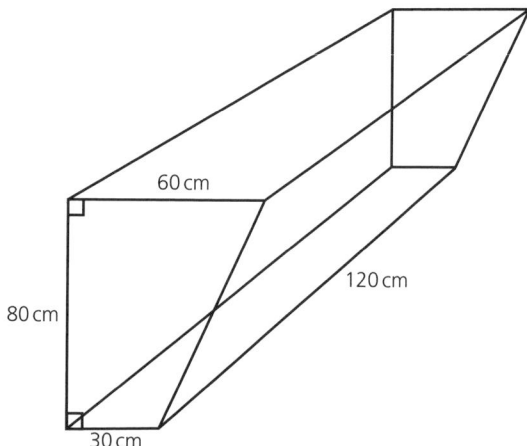

(ii) What is the depth of the water when the trough contains 40 litres?

(iii) The trough is being filled with a hose at a rate of 4 litres per minute.

At what rate is the depth increasing when there is 40 litres of water in the trough?

13 The diagram shows a water trough with a cross-section in the shape of an isosceles trapezium.

The sides AD and BC make an angle of 60° with the horizontal.

The trough is being filled by a large hose at a rate of 2 litres per second.

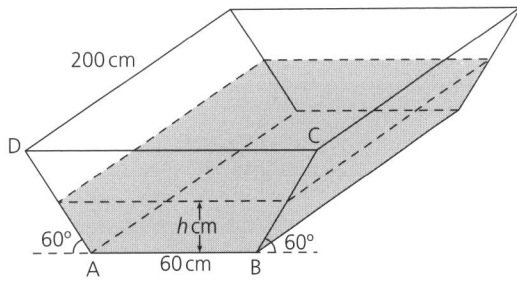

(i) Show that the volume of water, $V\,\text{cm}^3$, when it is $h\,\text{cm}$ deep is given by

$$V = 200h\left(60 + \frac{h}{\sqrt{3}}\right).$$

(ii) Hence find the rate at which the water level is rising when $h = 2\,\text{cm}$.

Stretch and challenge

1 The normal to the curve $g(x) = kx - x^2$ at $x = 2$ has a gradient of $\frac{1}{2}$.

    (i)   Find the value of k.

    (ii) The normal intersects the curve again at the point P.

        Find the co-ordinates of P.

2 The positions of a submarine (S) and a shipwreck (W) are shown in the diagram.

Due to territorial issues, the closest the submarine can get to the shipwreck is 30 km horizontally from its position at the bottom of the ocean.

The shipwreck is 15 km below the surface.

For the first 10 km in depth, the submarine can travel at 10 km/h but for the last 5 km, due to the increased pressure, it can only travel at 5 km/h.

Rather than go directly to W from S, the submarine goes to a point P on the line where the speed must change then changes course.

The distance S'P is x.

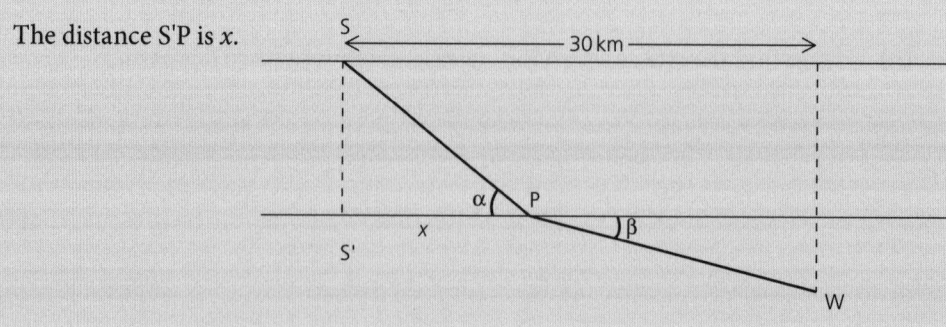

Stretch and challenge

(i) Find expressions for $\cos\alpha$ and $\cos\beta$ in terms of x.

(ii) Show that the minimum time to get from S to W occurs when $\cos\alpha = 2\cos\beta$.

(You do not need to show it is a minimum.)

Stretch and challenge

3 (i) A cylinder is enclosed in a sphere of radius R.

Find the dimensions r and h, in terms of R, of the cylinder with maximum volume.

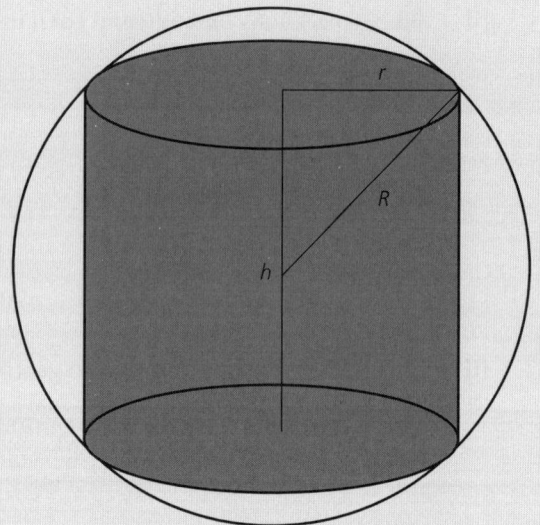

(ii) Find the ratio of the volume of the sphere to the cylinder with maximum volume in its simplest form.

4 The diagram shows the graph of $y = f'(x)$.

Sketch the original function, $y = f(x)$ on the graph, given that $f(0) = 4$ and $f'(2) = f(2)$.

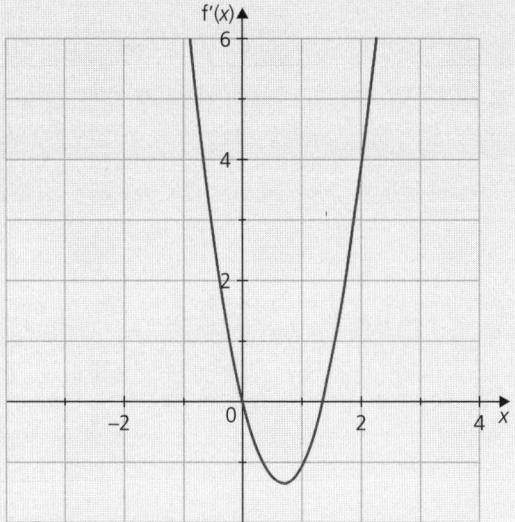

5 Three towns A, B and C form an isosceles triangle as shown in the diagram.

They are to be joined by three roads, DA, DB and DC.

Determine the position of the point D in the triangle so that the total length of the three roads is as small as possible.

(This point is known as the *Steiner Point*.)

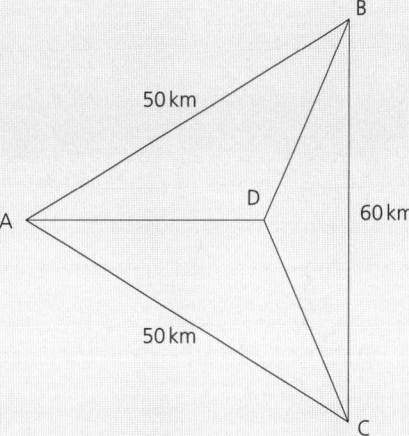

■ *Exam focus*

1 Find the co-ordinates of the point on the curve $y = \frac{1}{3}x^3 + \frac{1}{2}x^2 - 3x$ where the gradient is –1. [3]

2 Find the equation of the tangent and normal to the curve $f(x) = x + \frac{4}{x}$ at $x = 1$. [4]

3 Find the co-ordinates of the stationary points and determine their nature for the curve $y = x^4 - 8x^2 + 2$. [4]

4 Find the values of x for which the function $f(x) = 6x^2 - x^3$ is increasing. [3]

5 The ice cube shown in the diagram is melting so that the rate of decrease of the sides of the cube is 5 mm/minute.

Find the rate of decrease of the surface area of the ice cube when the side length is 2 cm. [4]

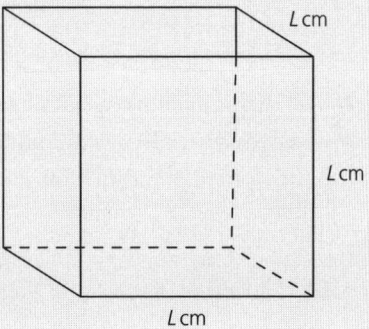

L cm

L cm

L cm

6 The equation of a curve is $y = \sqrt{1+6x}$.

A point with co-ordinates (x, y) moves along the curve in such a way that the rate of increase of x has the constant value 0.02 units per second.

Find the rate of increase of y at the instant when $x = 4$. [3]

7 The diagram shows a salt-shaker which consists of a cylinder radius r cm and height h cm and a hemi-sphere of radius r on top.

The salt-shaker has a volume of $100\,\text{cm}^3$.

(i) Show that $h - \dfrac{100}{\pi r^2} - \dfrac{2r}{3}$. [2]

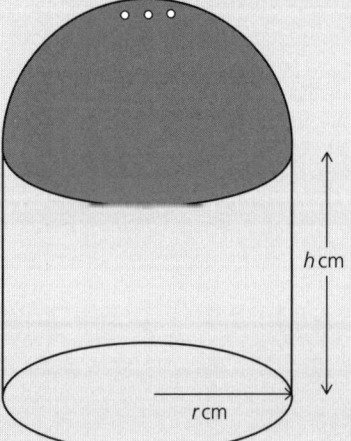

h cm

r cm

(ii) Show that the surface area, S, of the salt-shaker is given by

$$S = \frac{200}{r} + \frac{5\pi}{3}r^2$$ [3]

(iii) Given that r can vary, find the value of r for which S has a stationary value. [3]

(iv) Determine whether this stationary value from **(iii)** is a maximum or a minimum. [2]

6 Integration

Reversing differentiation

1 Find the following integrals.

(i) $\int \left(2x^3 + x\right) dx$

(ii) $\int (1 + 3x) \, dx$

(iii) $\int (2x - 5x^4) \, dx$

(iv) $\int \dfrac{4}{x^3} \, dx$

(v) $\int \left(\dfrac{x}{2} + 2x^5\right) dx$

(vi) $\int \left(\dfrac{1}{2x^2} + 6x\right) dx$

2 Find the following integrals.

(i) $\int 6\sqrt[3]{x} \, dx$

(ii) $\int x(2x - 3\sqrt{x}) \, dx$

(iii) $\int \left(\dfrac{2}{\sqrt{x}}\right) dx$

(iv) $\int 10\sqrt{x^3} \, dx$

3 Integrate $\int \dfrac{5x^2 - 1}{x^4}\, dx$

Hint: Divide first.

4 A curve is such that $\dfrac{dy}{dx} = \dfrac{3}{x^4} + 1$ and A(-1, 2) is a point on the curve.

Find the equation of the curve.

5 Find $f(x)$ if $f'(x) = 3\sqrt{x} - 2x$ and $f(4) = 30$.

6 A curve is such that $\dfrac{dy}{dx} = 1 - \dfrac{k}{x^2}$, where k is a constant.

The points M(1, 2) and N(−3, −6) are two points on the curve.

Find the equation of the curve.

What about the integral

$$\int \frac{1}{x}\,dx\,?$$

If we apply the rule we get

$$\int \frac{1}{x}\,dx = \int x^{-1}\,dx = \frac{x^0}{1} + c$$

The rule does not work here.

The answer to this integral is not required in this course.

Finding the area under a curve, Area as the limit of a sum, Areas below the *x* axis, The area between two curves, The area between a curve and the *y* axis

EXERCISE 6.2

1 Find the area between the line $y = 3x + 1$, the *x* axis and the lines $x = 1$ and $x = 2$

(i) using integration

(ii) using the formula for the area of a trapezium.

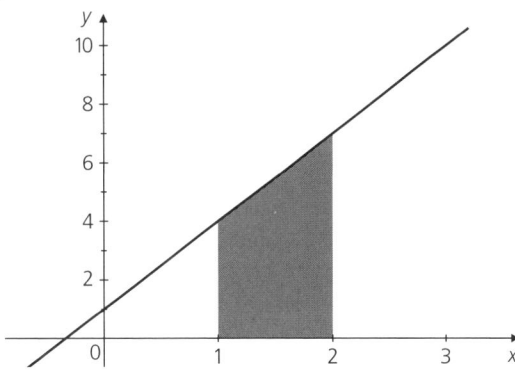

2 Find the area between the curve
$y = x^2 - 4x + 3$ and the *x* axis.

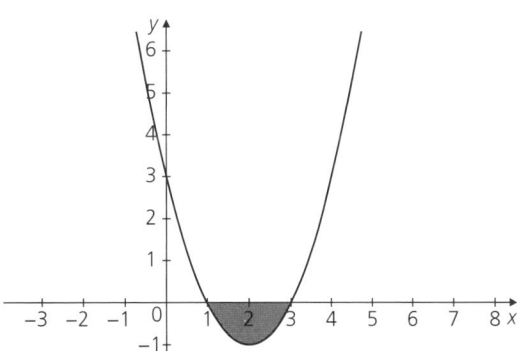

3 Find the area between the curve
$y = 2\sqrt{x}$, the y axis and the line $y = 4$.

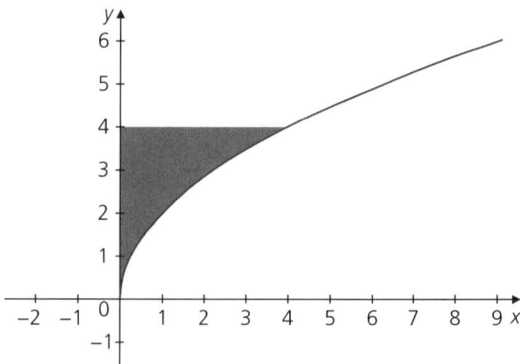

4 Find the area between the line $y = 2$
and the curve $y = 6 - x^2$.

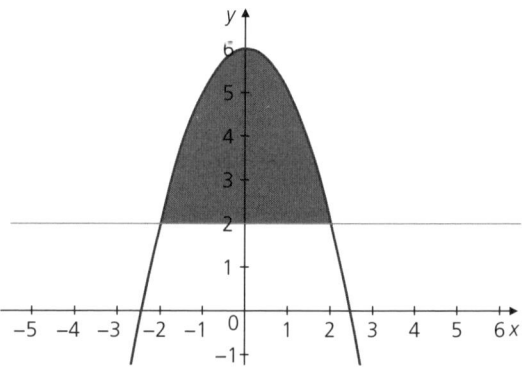

5 Find the area between the x axis, the curve $f(x) = 3x^2 + 1$ and the lines $x = -1$ and $x = 2$.

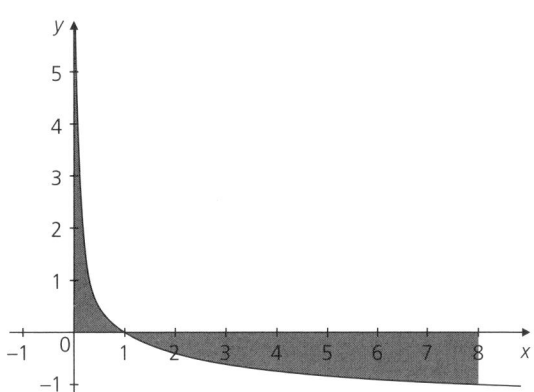

6 Find the area between the curve $y = \dfrac{2}{\sqrt[3]{x}} - 2$, the x axis and the lines $x = 0$ and $x = 8$.

7 Find the area between the line $y = x - 1$ and the curve $y = x^2 - 2x - 1$.

(It does not matter that part of the curve is below the x axis, the formula still works.)

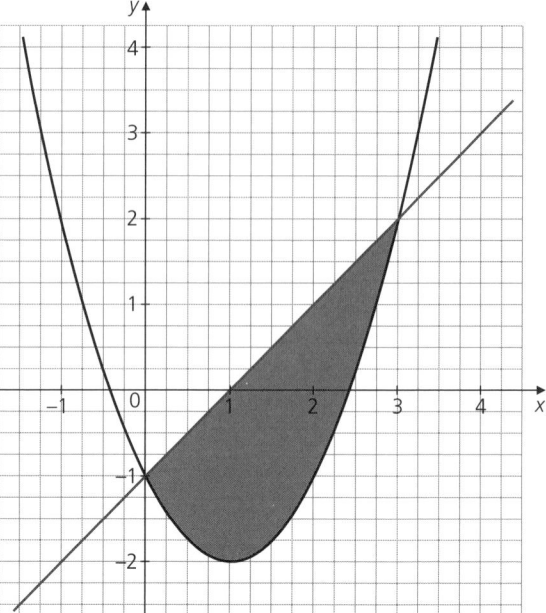

8 Find the area between the curves $y = 9 - x^2$ and $y = x^2 - 2x - 3$.

Start by drawing a sketch of the two graphs.

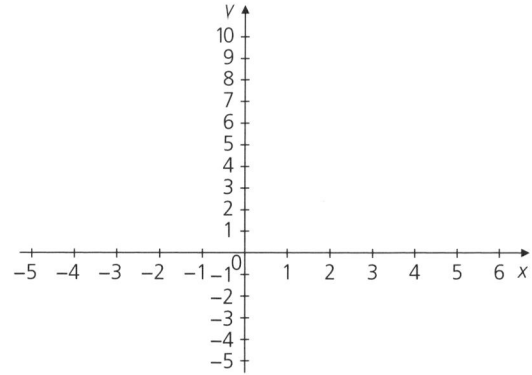

9 Find the area of the region between the curve $y = 2x^2 + 1$, the y axis and the lines $y = 3$ and $y = 9$.

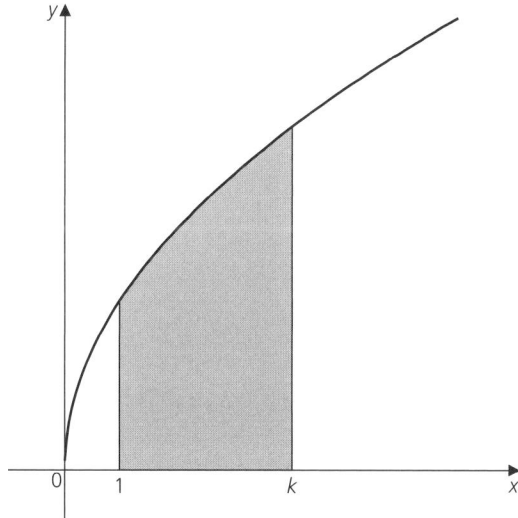

10 The area between the curve $y = 3\sqrt{x}$, the x axis and the lines $x = 1$ and $x = k$ is 14.

Find the value of k.

The reverse chain rule

EXERCISE 6.3

1 Find the following integrals.

(i) $\int (x-2)^3 \, dx$

(ii) $\int (3x+1)^5 \, dx$

(iii) $\int 2(1-6x)^9 \, dx$

(iv) $\int \left(\frac{x}{4}+3\right)^3 \, dx$

2 Find the integral $\int (2x-1)^{\frac{1}{2}} dx$, simplifying your answer as much as possible.

3 Find the following integrals.

(i) $\int \sqrt{1-x} \, dx$

(ii) $\int \frac{4}{(2x+1)^3} \, dx$

(iii) $\int \frac{2}{3(x-4)^5} \, dx$

(iv) $\int -\frac{14}{(7x+2)^6} \, dx$

(v) $\int \frac{3}{\sqrt{5-2x}} \, dx$

(vi) $\int \frac{12}{\sqrt[3]{1+\frac{x}{2}}} \, dx$

Improper integrals

1 Find $\int_0^1 \dfrac{1}{(2x+1)^3}\,\mathrm{d}x$.

2 Evaluate $\int_1^2 \dfrac{2}{\sqrt{x-1}}\,\mathrm{d}x$.

3 Evaluate $\int_0^\infty \dfrac{3}{(1+x)^2}\,\mathrm{d}x$.

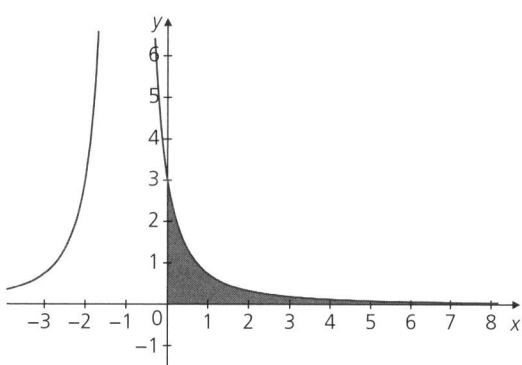

Finding volumes by integration

EXERCISE 6.5

1 (i) Find the volume of the solid formed when the area between the curve $y = \dfrac{4}{x}$ and the x axis for $x \geqslant 2$ is rotated 360° about the x axis.

Give your answer in terms of π and to 3 significant figures.

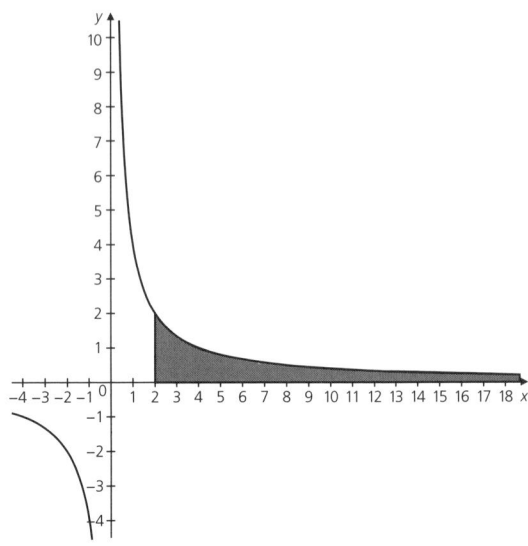

(ii) Find the volume of the solid formed when the area between the curve $y = \dfrac{4}{x}$, the x axis and the lines $x = 3$ and $x = 4$ is rotated completely around the x axis.

Give your answer in terms of π and to 3 significant figures.

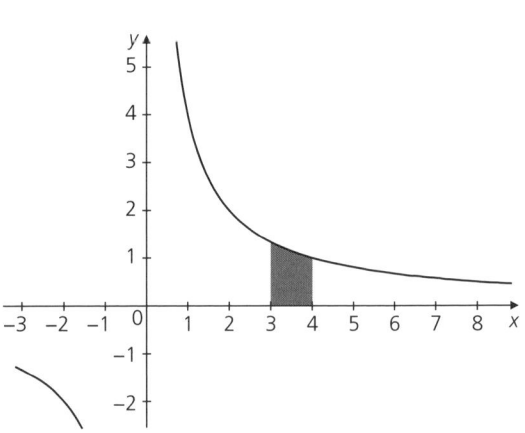

2 The area between the line $y = 3x$, the x axis and the lines $x = 0$ and $x = 2$ is rotated $360°$ about the x axis.

Find the volume of the solid generated.

Check that your answer is correct by finding the volume using the formula for the volume of a cone, $V = \frac{1}{3}\pi r^2 h$.

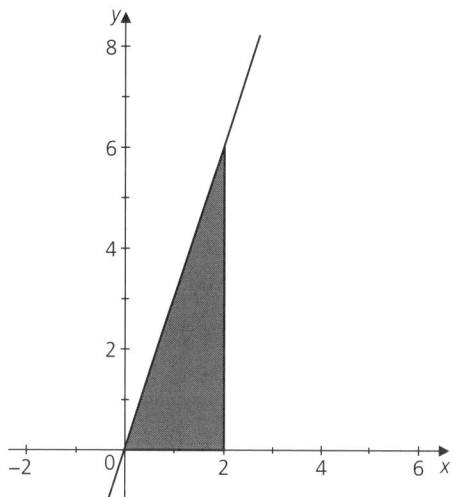

3 Find the volume of the solid formed when the shaded area under the curve $y = \dfrac{3}{2 - x}$ is rotated completely around the x axis.

Give your answer in terms of π.

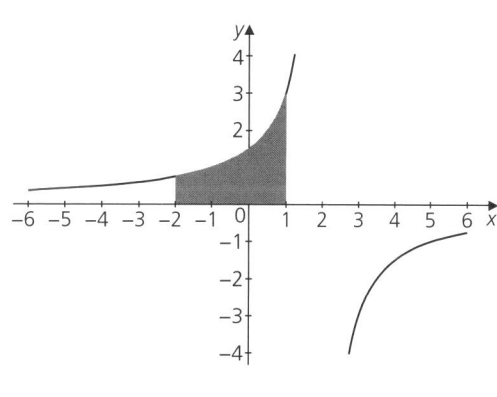

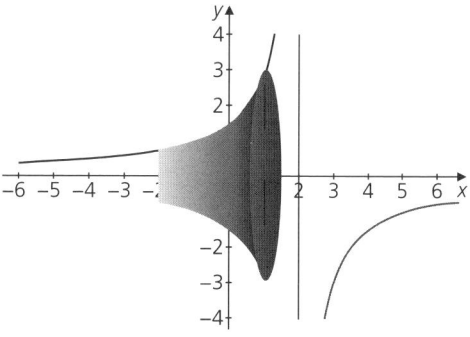

Integration

4 The part of the curve $y = 2(x^2 - 1)$ between $x = 1$, $x = 2$ and the x axis is rotated completely about the y axis.

Find the volume of the solid generated.

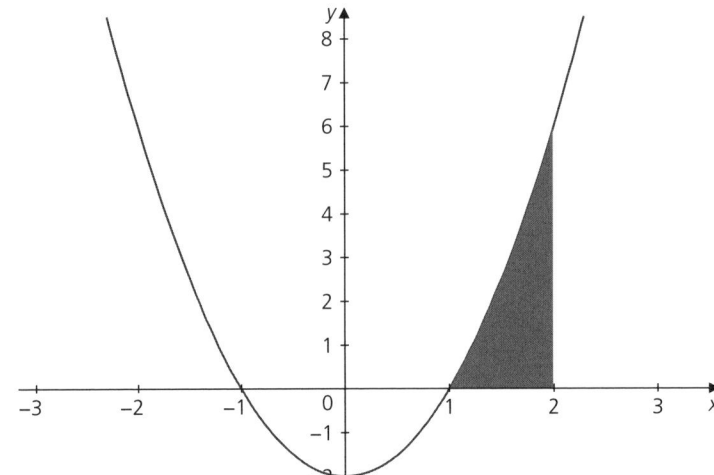

5 Show that the volume of the solid formed when the area between the curve $y = x(x-2)$ and the line $y = -\frac{1}{2}x$ is rotated 360° around the x axis is $\frac{27}{40}\pi$.

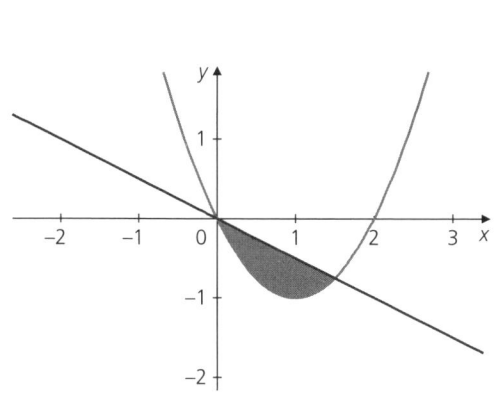

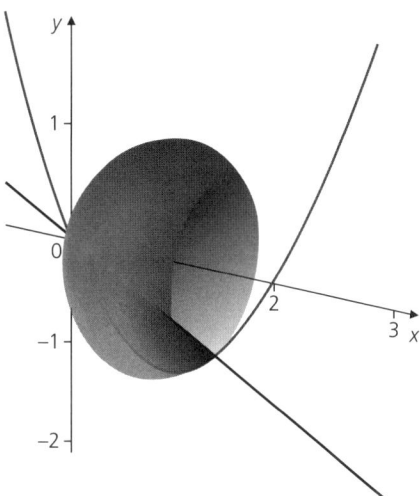

6 The diagram shows part of the curve $y = \dfrac{a}{x^2}$, where a is a positive constant.

Given that the volume obtained when the shaded region is rotated through $360°$ about the x axis is 21π, find the value of a.

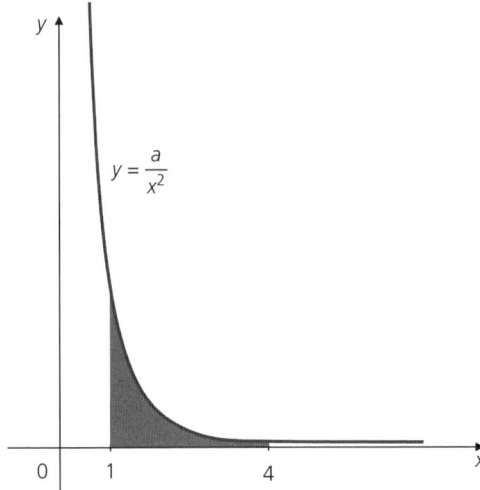

7 The diagram shows part of the curve $y = 4x + \dfrac{4}{x}$, which has a minimum point at M.

The line $y = 10$ intersects the curve at the points A and B.

(i) Find the co-ordinates of A, B and M.

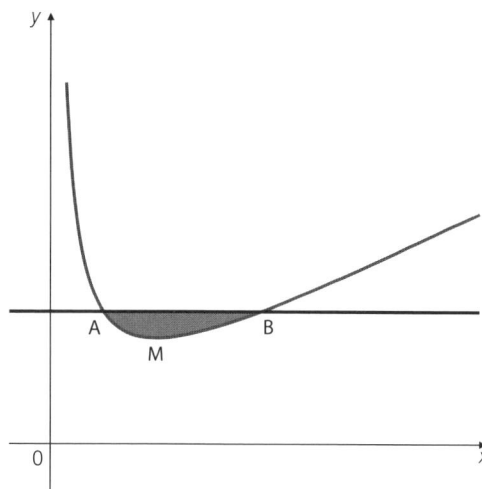

(ii) Find the volume obtained when the shaded region is rotated through 360° about the x axis.

Stretch and challenge

1 The shaded area between the curve $y = (x + 2)^2$ and the line $y = 2x + k$ between $x = -2$ and $x = 0$ is $\frac{10}{3}$.

Find the value of the constant k.

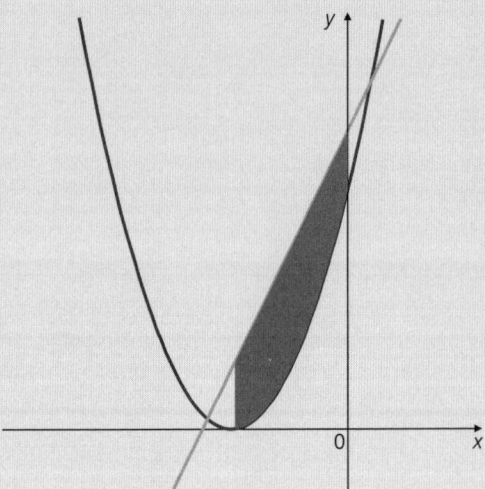

Stretch and challenge

2 The diagram shows a symmetrical plant pot with a circular base of radius
r centimetres, a circular top of radius R centimetres and straight sloping sides.

Show by integration that the volume of the pot is given by $\frac{1}{3}\pi h\left(R^2 + rR + r^2\right)\text{cm}^3$.

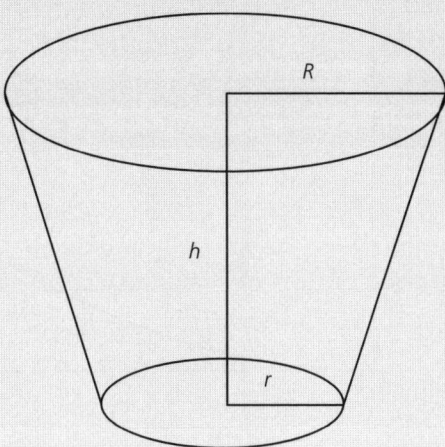

Stretch and challenge

3 The area under the function $f(x) = 2x^2$, $x > 0$, between $x = k$ and $x = k + 1$ is $\frac{109}{6}$.

Find the value of k.

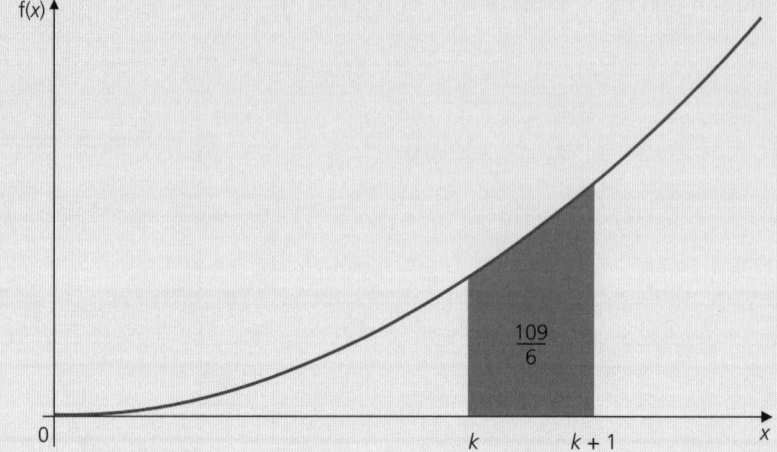

Stretch and challenge

4 A jeweller has a collection of solid gold rings for different-sized fingers.

The cross-section of each ring is a segment of a circle radius R as shown in the diagram. All the rings in the collection have the same width, w.

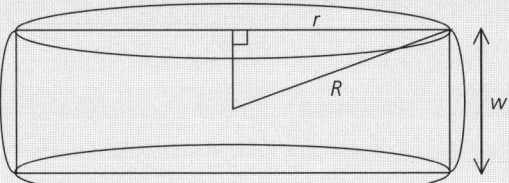

The jeweller tells a customer that, although the rings have different diameters, they all contain the same amount of gold.

Is this true?

Justify your answer.

Hint:

The equation of a circle with radius R is

$x^2 + y^2 = R^2$.

■ *Exam focus*

1 Find the equation of the curve through A(1, −2) for which $\dfrac{dy}{dx} = 2x + 1$. [3]

2 Find $\displaystyle\int \dfrac{3}{(4x+1)^2}\, dx$. [3]

3 Evaluate $\displaystyle\int_0^1 x^{-\frac{1}{2}}\, dx$. [3]

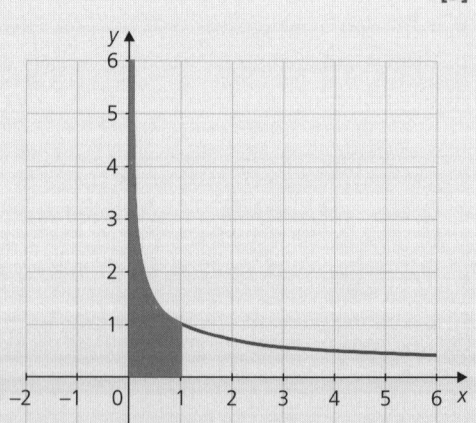

4 Evaluate $\int_{1}^{\infty} x^{-2} \mathrm{d}x$. [3]

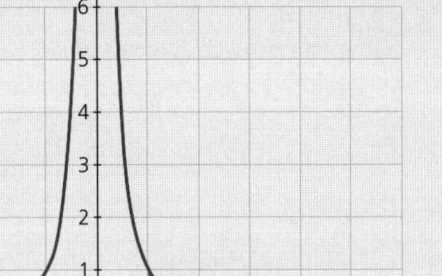

5 Find the area between the curve $y = 2\sqrt{x} + 1$, the y axis and the lines $y = 3$ and $y = 5$. [5]

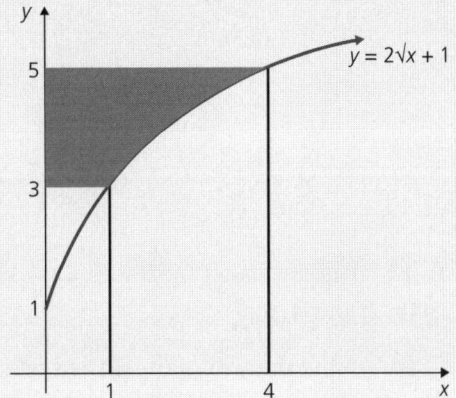

6 Find the area between the curves $y = 4\sqrt{x}$ and $y = x$. [4]

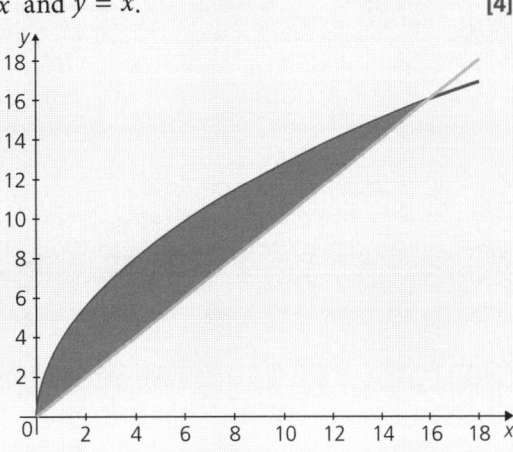

7 Find the volume of the solid formed when the shaded area under the curve $y = \dfrac{3}{2x+1}$ is rotated completely around the x axis.

Give your answer in terms of π.　　**[5]**

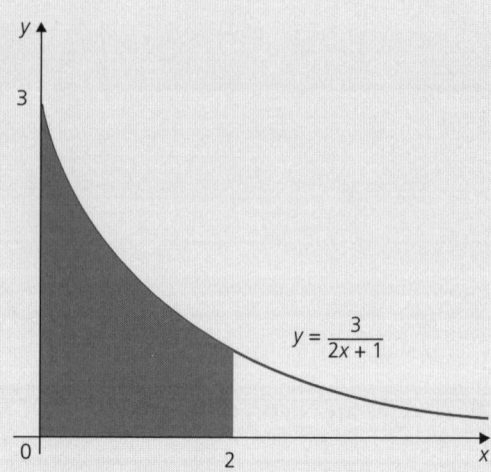

8 The diagram shows the line $y = 1 - x$ and the curve $y = \sqrt{(1 - x)}$ which intersect at $(0, 1)$ and $(1, 0)$.

(i)　Find the area of the shaded region.　　**[2]**

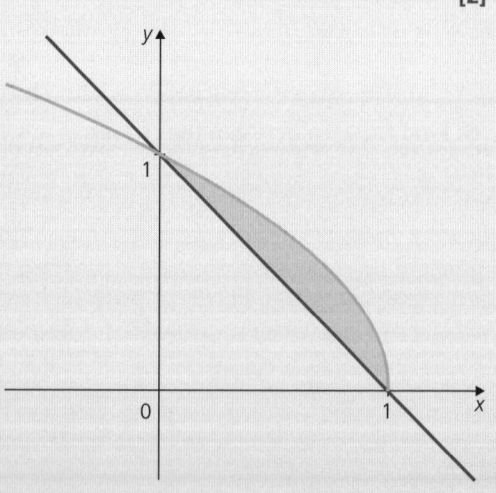

(ii) Find the volume obtained when the shaded region is rotated 360°

    (a) about the x axis   [5]

    (b) about the y axis.   [5]

9 Find the volume of the solid formed when the area between the curve $y = x^3$ and the line $y = x$ is rotated 360°

(i) about the x axis [5]

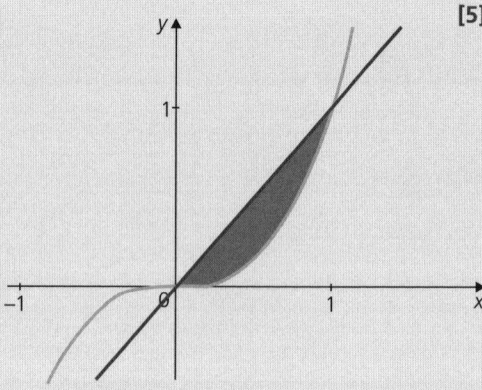

(ii) about the y axis. [5]

7 Trigonometry

Trigonometry background, Trigonometrical functions

1 Find the lengths and angles marked.

(i)

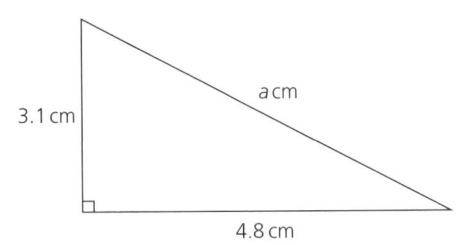

3.1 cm

a cm

4.8 cm

(ii)

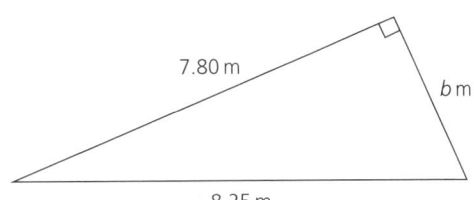

7.80 m

b m

8.25 m

(iii)

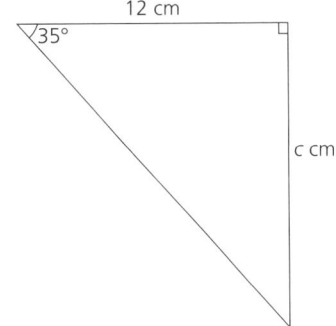

12 cm

35°

c cm

(iv)

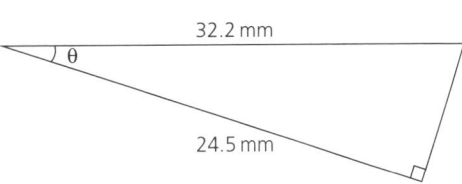

32.2 mm

θ

24.5 mm

2 Find the lengths and angles marked.

(i)

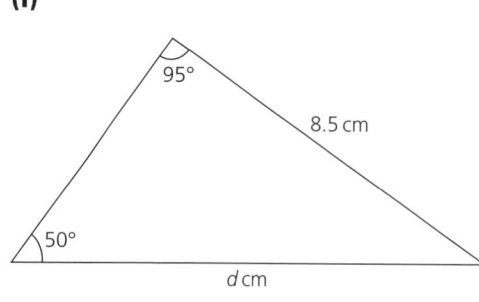

(ii)

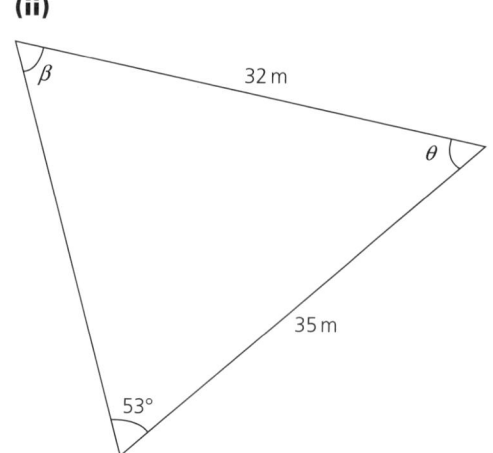

(iii)

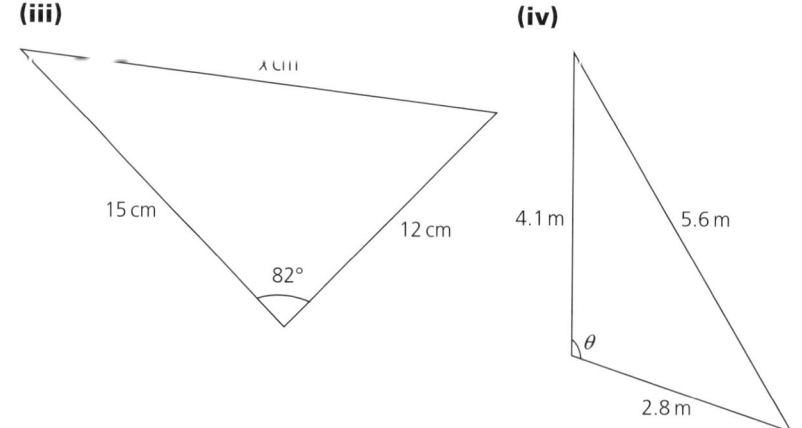

(iv)

3 Find the areas of the triangles in question **2**.

(i)

(ii)

(iii)

(iv)

4 Find the unknown lengths marked in the following triangles.

(i)

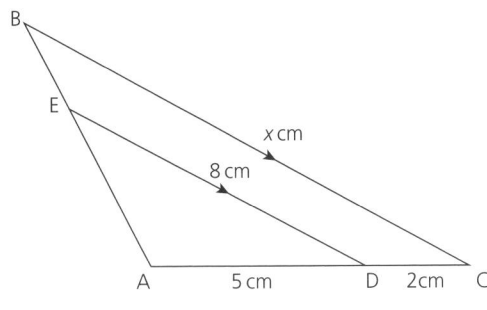

(ii)

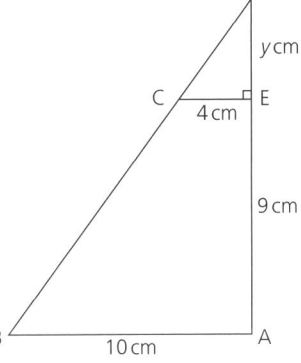

Trigonometrical functions for angles of any size

1 Use the unit circle to complete the table with angles and exact values.

All angles are between 0° and 360°.

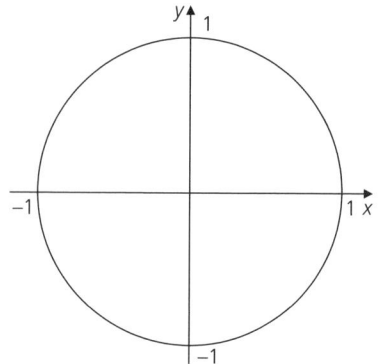

Angle	Other angle that has the same value	Exact value
sin 30°	sin ....°	
sin 210°	sin ....°	
cos ....°	cos ....°	$\dfrac{1}{2}$
tan 30°	tan ....°	
cos 150°	cos ....°	
tan ....°	tan ....°	$-\sqrt{3}$

2 Prove the following identities.

(i) $\sin x \tan x \equiv \dfrac{1}{\cos x} - \cos x$

(ii) $\tan x (1 - \sin^2 x) \equiv \sin x \cos x$

(iii) $\dfrac{1 + \cos x}{\sin x} \equiv \dfrac{\sin x}{1 - \cos x}$

(iv) $\tan^2 x - \sin^2 x \equiv \tan^2 x \sin^2 x$

3 Prove the following identities.

(i) $\sin^4 x - \cos^4 x \equiv \sin^2 x - \cos^2 x$

(ii) $\tan x + \dfrac{1}{\cos x} \equiv \dfrac{\cos x}{1 - \sin x}$

(iii) $\dfrac{1}{1 - \cos x} + \dfrac{1}{1 + \cos x} \equiv \dfrac{2}{\sin^2 x}$

(iv) $\dfrac{1 + \tan^2 x}{1 - \tan^2 x} \equiv \dfrac{1}{2\cos^2 x - 1}$

4 Prove the following identities.

(i) $\dfrac{\cos \theta}{\tan \theta (1 + \sin \theta)} \equiv \dfrac{1}{\sin \theta} - 1$

(ii) $\dfrac{\cos x}{1 - \cos x} - \dfrac{\cos x}{1 + \cos x} \equiv \dfrac{2}{\tan^2 x}$

EXERCISE 7.3

1 Given that $x = \cos^{-1}\left(\frac{1}{4}\right)$ and x is an acute angle, find the exact value of

    (i) $\sin x$ **(ii)** $\tan^2 x$.

2 Given that $\tan x = -\frac{2}{7}$ and $90^{\circ} \leqslant x \leqslant 180^{\circ}$, find the exact value of

    (i) $\cos x$ **(ii)** $\sin^2 x$.

3 Solve the following equations where $0^{\circ} \leqslant x \leqslant 360^{\circ}$.

    (i) $2\sin x = -1$ **(ii)** $\cos 2x = \frac{1}{2}$

    (iii) $\tan x + 1 = 3$ **(iv)** $\sin(x - 30^{\circ}) = 1$

4 Solve $2\cos 3x - \sin 3x = 0$ for $0^{\circ} \leqslant x \leqslant 180^{\circ}$.

5 Show on the unit circle, the two angles between $0°$ and $360°$ that satisfy the given equations.

Write down the angles.

(i) $\sin x = \dfrac{1}{4}$

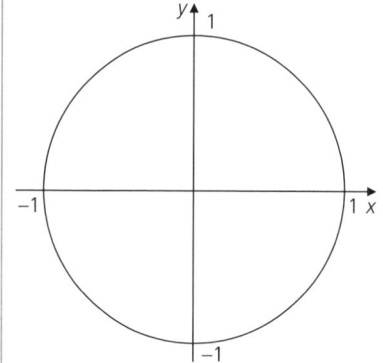

(ii) $\cos x = -0.7$

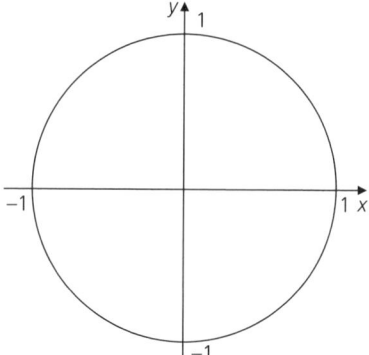

(iii) $\tan x = \dfrac{7}{8}$

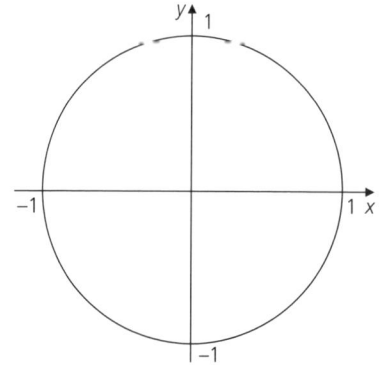

(iv) $\sin x = -\dfrac{2}{5}$

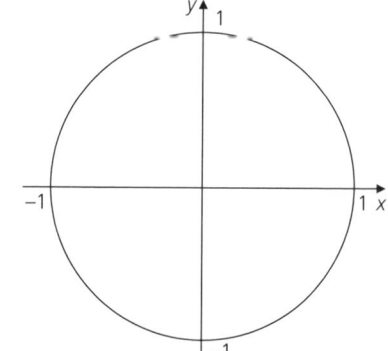

6 (i) Prove the identity $\left(\dfrac{1}{\cos\theta} + \tan\theta\right)^2 \equiv \dfrac{1+\sin\theta}{1-\sin\theta}$.

(ii) Hence solve the equation $\left(\dfrac{1}{\cos\theta} + \tan\theta\right)^2 = \dfrac{3}{7}$ for $0° \leqslant \theta \leqslant 360°$.

7 Solve the following equations over the given intervals.

Use the unit circle diagram to help you find all the solutions.

(i) $2\cos^2 x = \cos x$ for $0° \leqslant x \leqslant 360°$

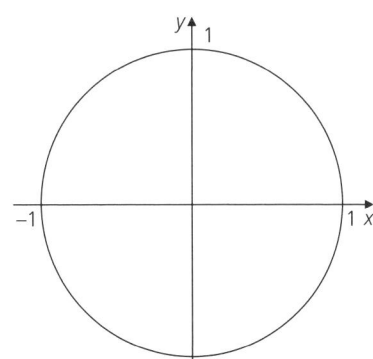

(ii) $4 + 5\cos x = 2\sin^2 x$ for $-180° \leqslant x \leqslant 180°$

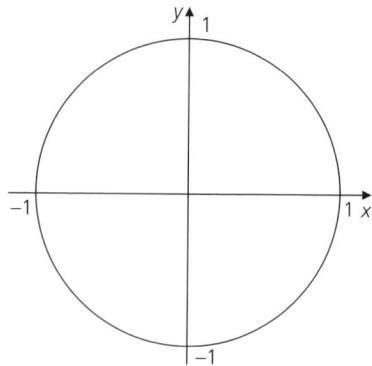

(iii) $\sin^2 x = 3(1 + \cos x)$ for $0° \leqslant x \leqslant 360°$

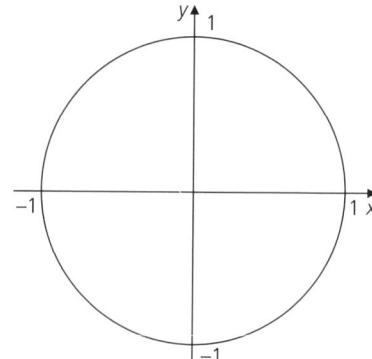

8 (i) Given that $4\cos^2 x + 7\sin x - 2 = 0$

show that, for real values of x, $\sin x = -\dfrac{1}{4}$.

(ii) Hence solve the equation

$4\cos^2(\theta - 20°) + 7\sin(\theta - 20°) - 2 = 0$ for $0° \leqslant \theta \leqslant 360°$.

Circular measure

EXERCISE 7.4

1 Complete the table.

Give all your answers as exact fractions of π in the radians column.

Degrees	Radians
120°	
	$\dfrac{\pi}{5}$
330°	
	$\dfrac{5\pi}{6}$
240°	
	$\dfrac{3\pi}{10}$
225°	
	$\dfrac{7\pi}{4}$

2 Complete the table.

Give the answers in degrees to 1 decimal place and those in radians to 3 significant figures.

Degrees	Radians
12°	
	0.190
145°	
	2.95
235°	
	5.04
342.5°	
	1.37

3 Use the unit circle to complete the table with angles and exact values.

All angles are between 0 and 2π radians.

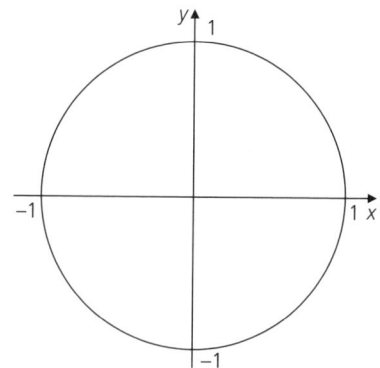

Angle	Other angle that has the same value	Exact value
$\sin \dfrac{\pi}{4}$	sin	
cos	cos	$\dfrac{\sqrt{3}}{2}$
$\tan \dfrac{7\pi}{6}$	tan	
$\cos \dfrac{3\pi}{4}$	cos	
$\sin \dfrac{5\pi}{3}$	sin	
tan	tan	1

4 Solve the following equations over the given intervals.

Use the unit circle diagram to help you find all the solutions.

(i) $2\tan x = \tan^2 x$ for $0 \leqslant x \leqslant 2\pi$

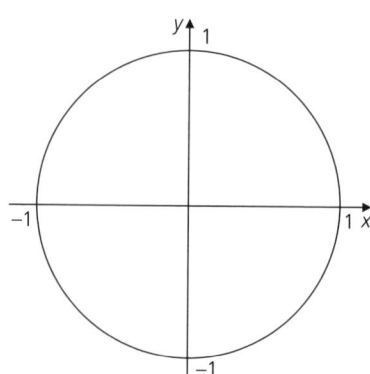

(ii) $4\sin^2 x + 4\sin x - 3 = 0$ for $0 \le x \le 2\pi$

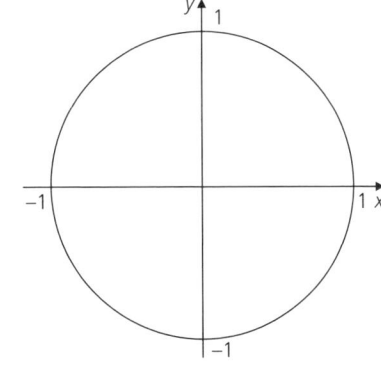

(iii) $\cos x = -3\tan x$ for $-\pi \le x \le \pi$

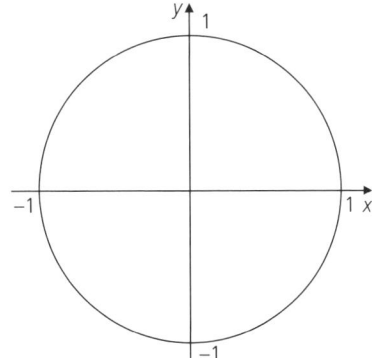

(iv) $2\sin^2 x = 3 - 3\cos x$ for $0 \le x \le 2\pi$

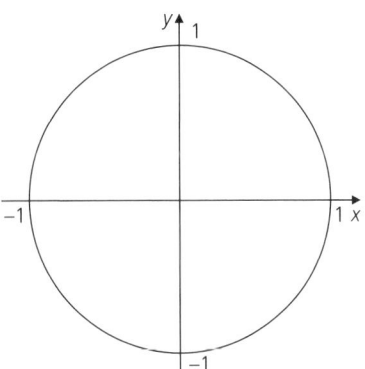

(v) $4\cos^2 x = 3\sin^2 x - \sin x \cos x$ for $-\pi \leqslant x \leqslant \pi$

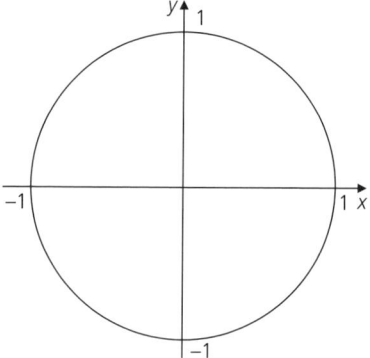

(vi) $3\sin^2 x + 5\cos^2 x = 9\sin x$ for $0 \leqslant x \leqslant 2\pi$

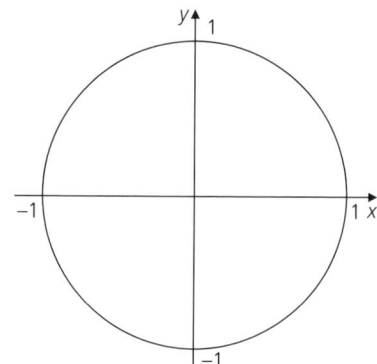

The length of an arc of a circle, The area of a sector of a circle

EXERCISE 7.5

1 Find the perimeter and area of the shaded regions.

(i)

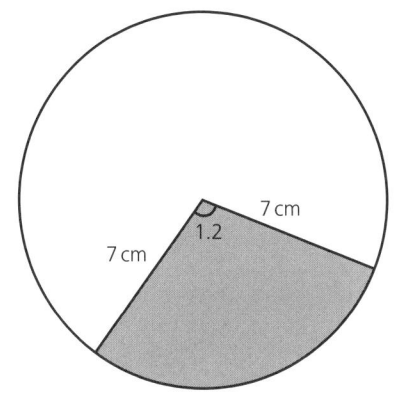

7 cm

7 cm

1.2

(ii)

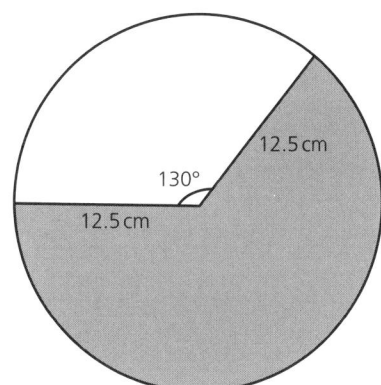

130°

12.5 cm

12.5 cm

2 Find the value of θ in radians and degrees.

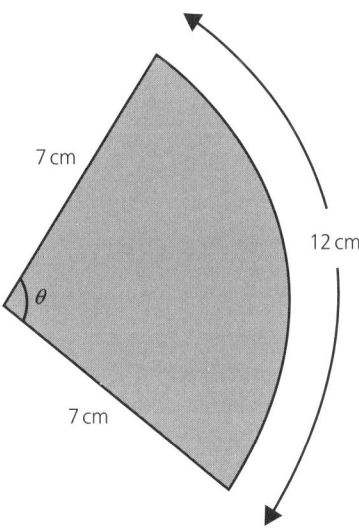

7 cm

7 cm

12 cm

θ

3 The diagram shows a circle centre O and radius 5 cm.

The tangents at A and B meet at the point C.

Angle AOB is $\dfrac{2\pi}{3}$ radians.

Find the perimeter and area of the shaded region.

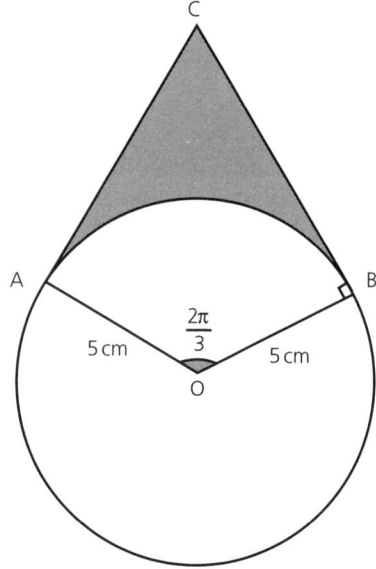

4 In the diagram, OQR is a sector of a circle with centre O and radius 10 cm.

Angle QOR = 0.8 radians and RP is the perpendicular from R to OQ.

(i) Find the perimeter of the shaded region.

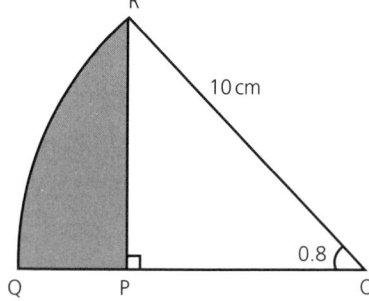

(ii) Find the area of the shaded region.

5 The diagram shows a sector of a circle of radius 6 cm.

(i) If the shaded area is 36 cm², show that
$\theta = \sin\theta + 2$.

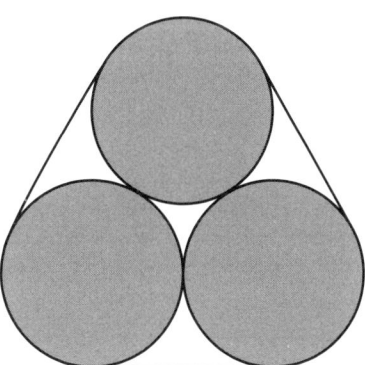

6 cm

6 cm

θ

(ii) Given that the perimeter of the sector is 27.3 cm, find the value of θ.

6 The diagram shows the cross-section of three pencils, each of radius 4 cm, with a rubber band fitted around them.

Find the length of the rubber band.

7 The diagram shows a circle with centre O and radius 8 cm.

The line AC is a tangent to the circle.

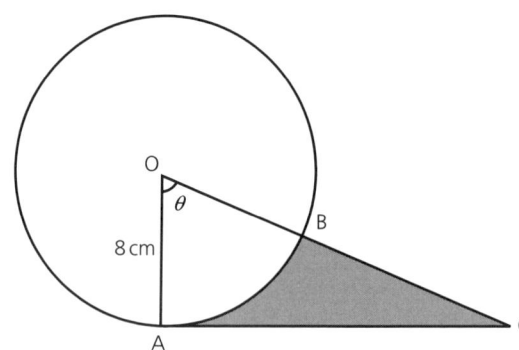

(i) Show that the area of the shaded region is $32(\tan\theta - \theta)\ \text{cm}^2$.

(ii) When $\theta = \dfrac{\pi}{3}$ find the perimeter of the shaded region.

8 The diagram shows a sector OAB of a circle, centre O and radius 16 cm.

The mid-points of OB and OA are D and C respectively.

The length of DC is 8 cm.

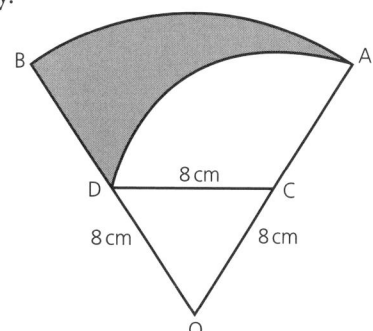

AD is an arc of the circle, centre C and radius 8 cm.

The shaded region is bounded by the line BD and the arcs AB and AD.

(i) Show that the angle ACD $= \frac{2}{3}\pi$ radians.

(ii) Show that the perimeter of the shaded region is $\left(\frac{32}{3}\pi + 8\right)$ cm.

(iii) Find the exact area of the shaded region.

Other trigonometrical functions

EXERCISE 7.6

1 Sketch the graphs.

(i) $y = \sin 2x + 1$ $0° \leqslant x \leqslant 360°$

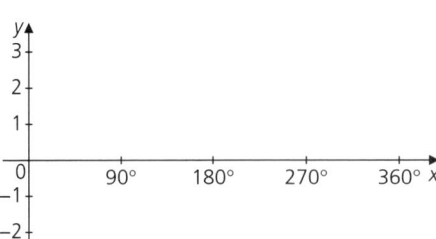

(ii) $y = 3\cos x - 2$ $0 \leqslant x \leqslant 2\pi$

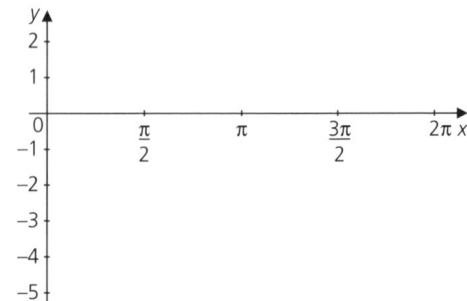

(iii) $y = \tan 2x$ $0° \leqslant x \leqslant 360°$

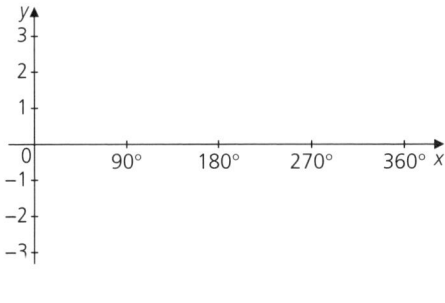

(iv) $y = 1 + 3\sin\dfrac{x}{2}$ $0 \leqslant x \leqslant 2\pi$

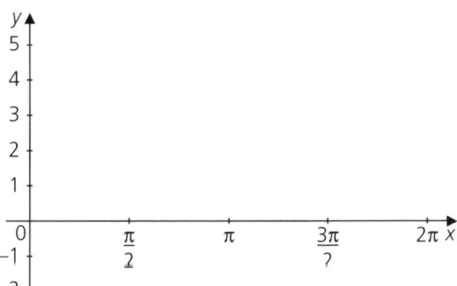

(v) $y = 2 - 2\cos 3x$ $0° \leqslant x \leqslant 360°$

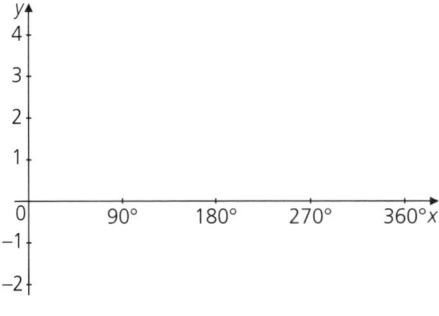

(vi) $y = 4 - 3\sin 2x$ $0 \leqslant x \leqslant 2\pi$

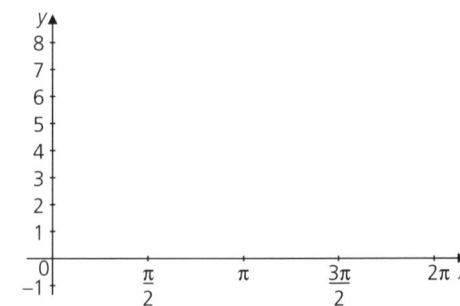

2 The diagram shows the graph of $y = A \sin Bx + C$.

Write down the values of A, B and C.

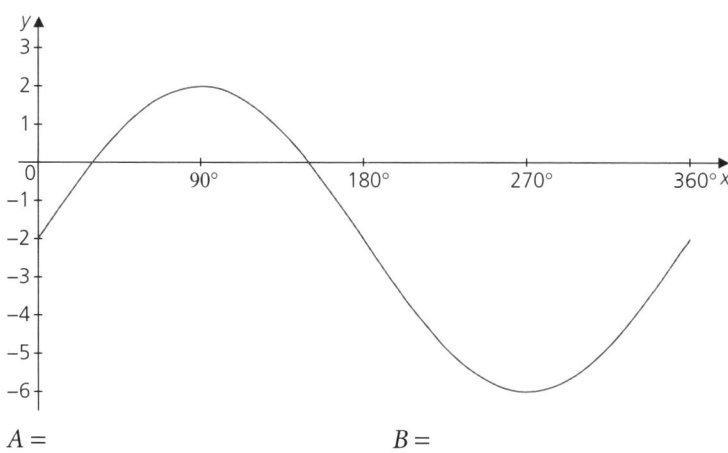

$A =$ $B =$ $C =$

3 The diagram shows the graph of $y = A \cos Bx + C$.

Write down the values of A, B and C.

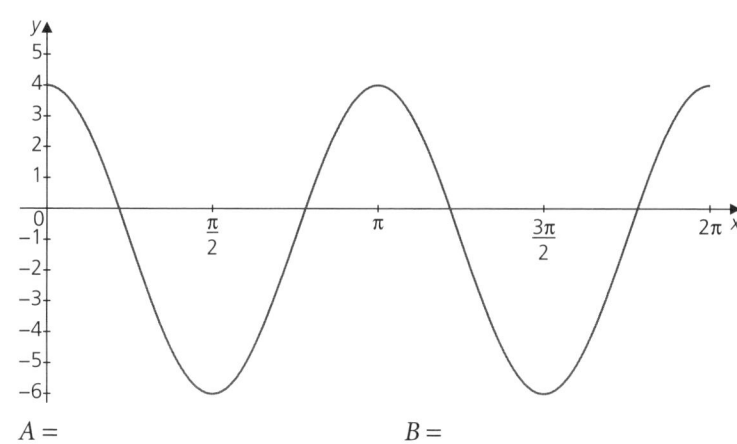

$A =$ $B =$ $C =$

4 The diagram shows the graph of $y = A \sin Bx + C$.

Write down the values of A, B and C.

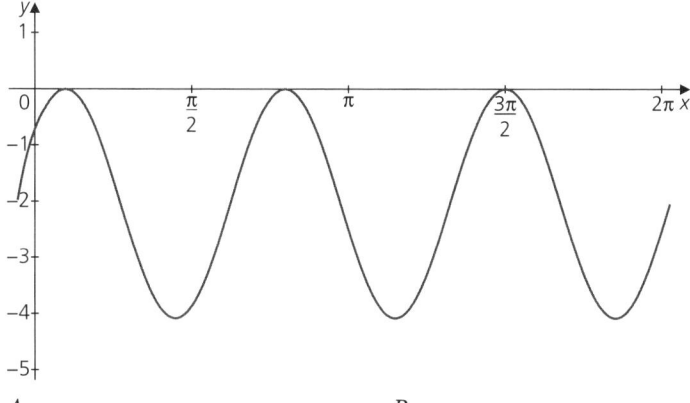

$A =$ $B =$ $C =$

5 The diagram shows the graph of $y = A \cos Bx + C$.

Write down the values of A, B and C.

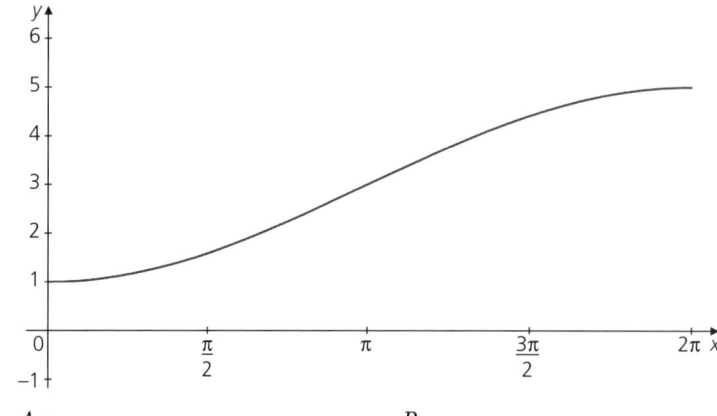

$A =$ $B =$ $C =$

6 The diagram shows the graph of $y = A - B \sin Cx$.

Write down the values of A, B and C.

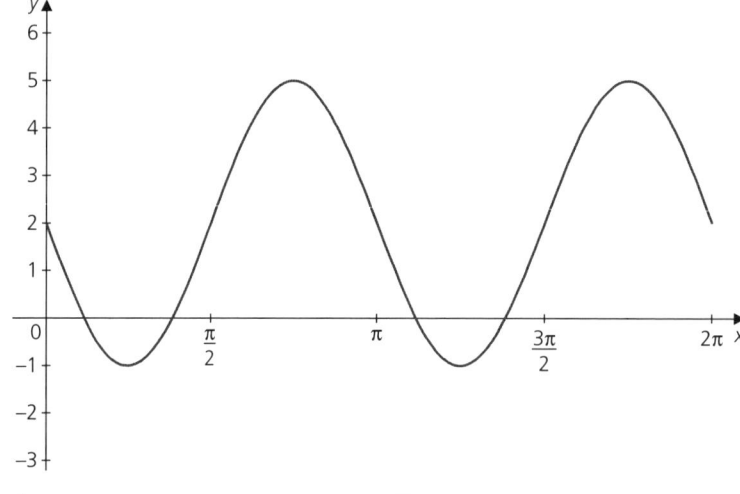

$A =$ $B =$ $C =$

7 The diagram shows the graph of $y = \tan Ax + B$.

Write down the values of A and B.

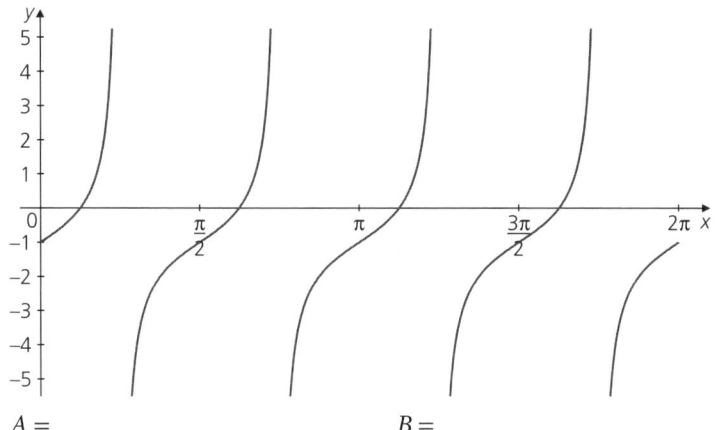

$A =$ $B =$

8 The temperature of the water at a lake (T) alternates regularly throughout the day.

At t hours after 12 noon the temperature is given by the equation

$T = 2\cos(3\pi t) - 1$.

(i) Find the period of T.

(ii) Sketch the graph of T between 12 noon and 4 pm.

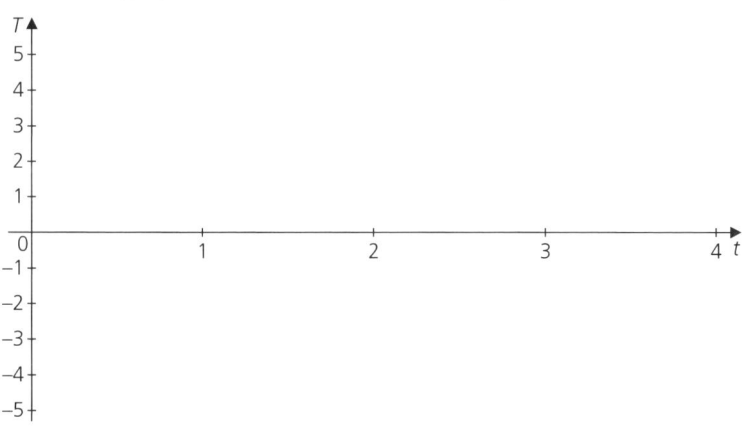

(iii) From the graph, find the warmest temperature of the water, and the times it reaches this temperature between 12 noon and 4 pm.

9 (i) Sketch the curve $y = 3\sin x$ for $0 \leqslant x \leqslant 2\pi$.

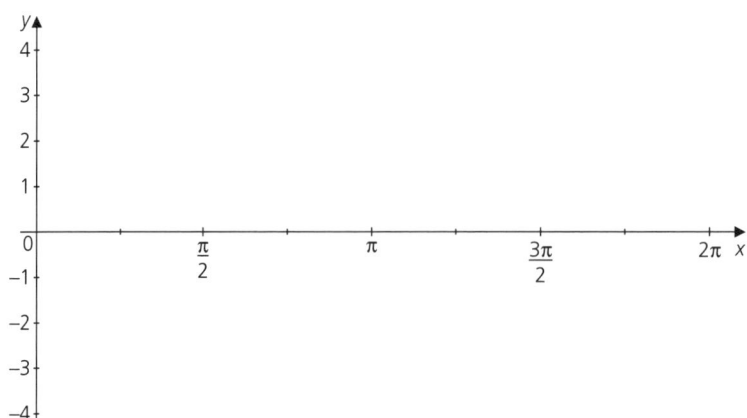

(ii) By adding a suitable straight line to your sketch, determine the number of real roots of the equation $3\pi \sin x = \pi + x$.

State the equation of the straight line.

10 A function f is defined by f: $x \mapsto 2 - 3\tan\left(\frac{1}{2}x\right)$ for $0 \leqslant x \leqslant \pi$.

(i) State the range of f.

(ii) State the exact value of $f\left(\frac{1}{3}\pi\right)$.

(iii) Sketch the graph of $y = f(x)$.

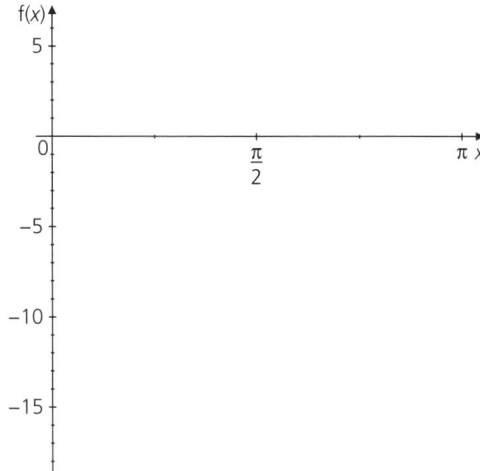

(iv) Obtain an expression, in terms of x, for $f^{-1}(x)$.

11 The function f is defined by $f : x \mapsto a - b \sin x$, where a and b are both positive constants.

    (i) The minimum value of f is -2 and the maximum value is 8.

        Find the values of a and b.

(ii) Hence sketch $y = f(x)$ for $0 \leqslant x \leqslant 2\pi$.

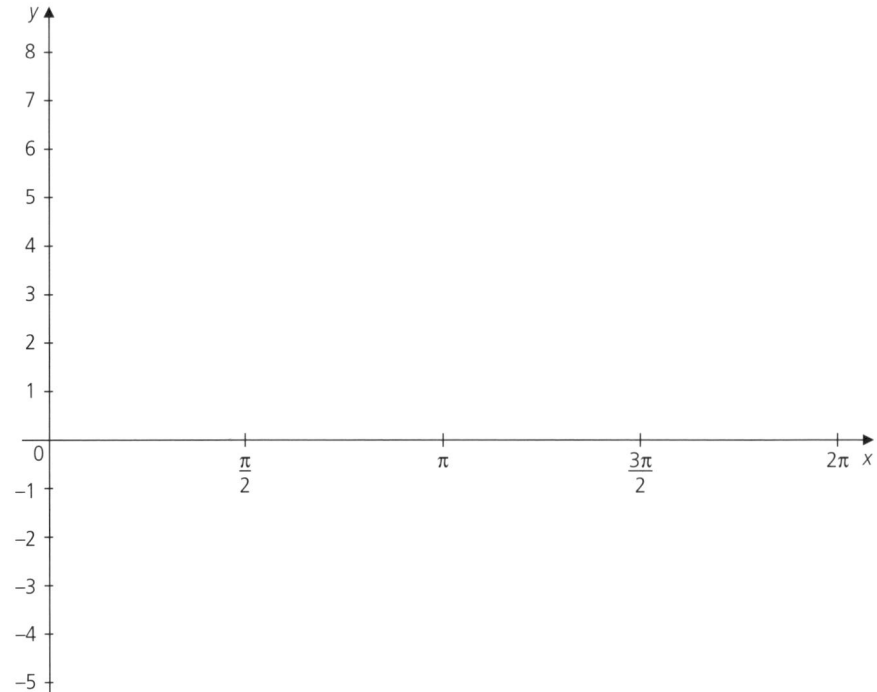

12 The acute angle x radians is such that $\cos x = k$, where k is a positive constant and $0 \leqslant x \leqslant \dfrac{\pi}{2}$.

Express the following in terms of k.

(i) $\cos(-x)$ $=$

(ii) $\cos(\pi - x)$ $=$

(iii) $\sin x$ $=$

(iv) $\cos(\pi + x)$ $=$

(v) $\tan x$ $=$

13 The obtuse angle x radians is such that $\tan x = -k$, where k is a positive constant and $\dfrac{\pi}{2} \leqslant x \leqslant \pi$.

Express the following in terms of k.

(i) $\tan(-x)$ $=$

(ii) $\tan(\pi - x)$ $=$

(iii) $\cos x$ $=$

(iv) $\sin(\pi + x)$ $=$

(v) $\tan\left(\tfrac{1}{2}\pi + x\right) =$

14 Solve the following equations for $0 \leqslant x \leqslant 2\pi$.

Give your answers in radians.

(i) $2\sin x = -\sqrt{3}$

(ii) $\cos 3x = -1$

(iii) $2\tan x - 1 = 1$

(iv) $\frac{1}{2}\sin(2x+1) = 0.1$

15 The height of a carriage above the ground (h) on a Ferris Wheel ride after t seconds is given by the equation:

$$h = 54 + 53\sin\left(\frac{\pi}{20}t\right).$$

(i) What is the period of the function (i.e. how long does one complete revolution of the wheel take)?

(ii) Find the maximum height of the ride.

(iii) Find how long it takes after the start of the ride to get to a height of 30 m off the ground.

Stretch and challenge

1 Solve the equation $3 \sin 3\theta + 3 = 2 \cos^2 3\theta$ for $-180° \leqslant \theta \leqslant 180°$.

2 The height H of the sea above sea level at a certain jetty is given by the equation

$H = 2 \sin \frac{\pi}{2}t + 3$, where t is the time in hours from midnight.

(i) Sketch the graph of $H = 2 \sin \frac{\pi}{2}t + 3$ for $0 \leqslant t \leqslant 8$.

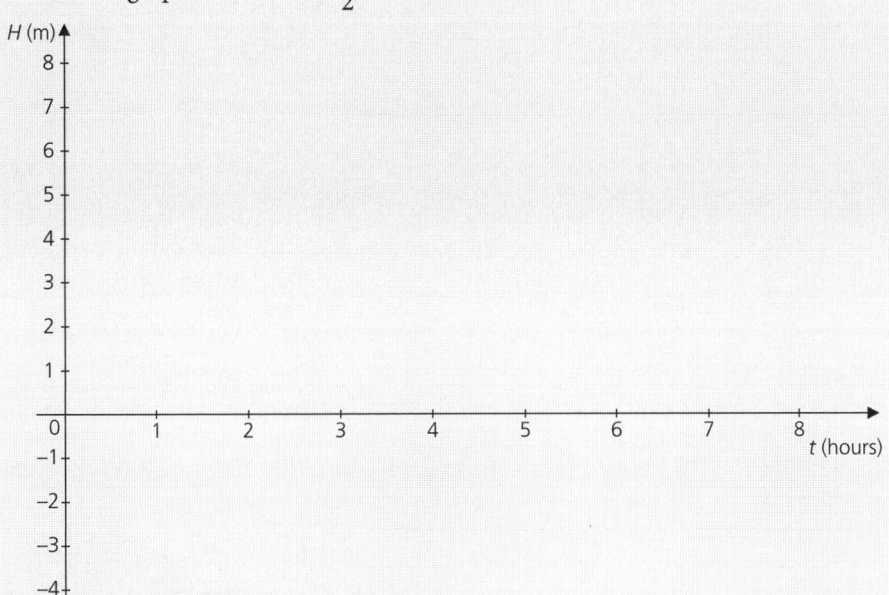

(ii) Find the period of H.

(iii) At what time will the first high tide occur?

Stretch and challenge

3 The height of water, h, above a reef is modelled by the equation

$$h = 9 - 3\cos\left(\frac{\pi}{8}t\right)$$

where t is time in hours after low tide.

A ship which has run aground can only be refloated when the water level is 10.5 m above the reef.

The refloating can only happen after a low tide which is at 9.30 am the next morning.

Between what times would it next be possible to refloat the ship?

4 A solar eclipse occurs when the moon passes between the Earth and the Sun.

Assume that the radius of both the Sun and the Moon when seen from the Earth is the same.

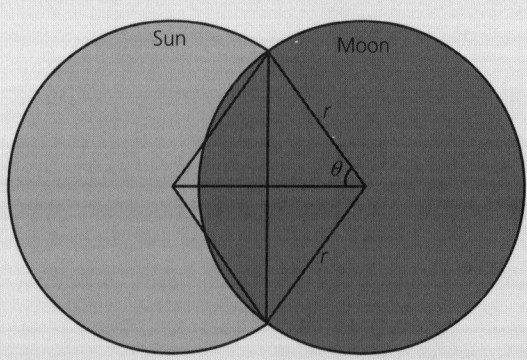

Stretch and challenge

(i) Show that, when 50% of the area of the sun is covered, $2\sin 2\theta = 4\theta - \pi$.

(ii) Find the angle θ when the distance between the centres of the Sun and the Moon is the same as the radius r.

(iii) Hence find the percentage area of the sun that is covered when the distance between the centres is the same as the radius r.

■ *Exam focus*

1 Find the value of the constants in the equations of the following graphs.

(i) $y = a \sin bx$ [2]

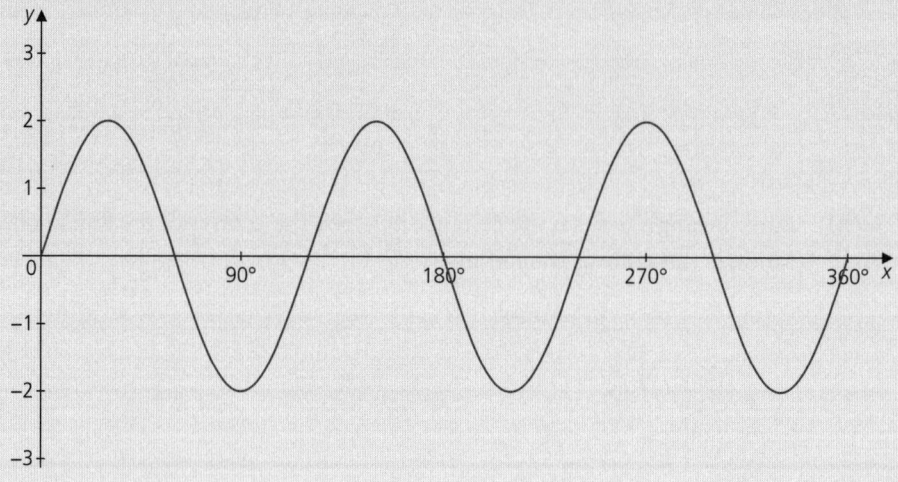

(ii) $y = a - b \cos cx$ [3]

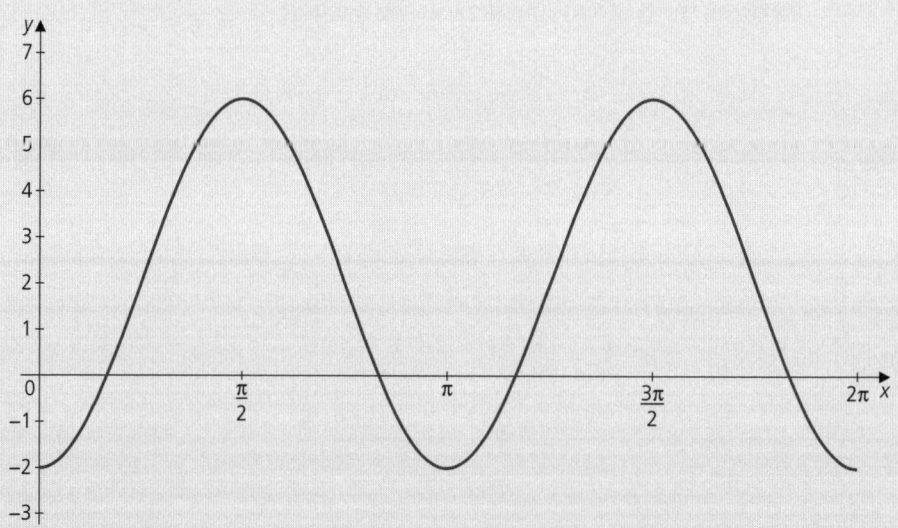

2 The acute angle x radians is such that $\sin x = k$, where k is a positive constant.

Express, in terms of k

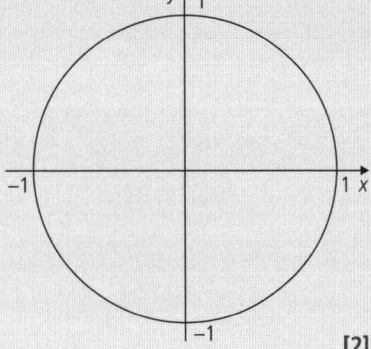

(i) $\sin(\pi - x)$ [1]

(ii) $\cos x$ [2]

(iii) $\tan\left(\dfrac{\pi}{2} - x\right)$ [2]

3 Prove the identity $1 + \tan^2\theta \equiv \dfrac{1}{\cos^2\theta}$. [4]

4 Prove the identity $\tan x + \dfrac{1}{\tan x} \equiv \dfrac{1}{\sin x \cos x}$ [3]

5 Prove the identity $\dfrac{1-\cos x}{\sin x} + \dfrac{\sin x}{1-\cos x} \equiv \dfrac{2}{\sin x}$ [4]

6 Solve the equation $2\cos 2x = \sqrt{3}$ for $0° \leqslant x \leqslant 360°$ [4]

7 Solve the equation $\sqrt{3}\sin 2x + \cos 2x = 0$ for $-\pi \leqslant x \leqslant \pi$. [4]

8 Solve the equation $\cos^2 x - 1 = \sin x$ for $-\pi \leqslant x \leqslant \pi$. [4]

9 Solve the equation $2\sin x \tan x = 3$ for $-\pi \leqslant x \leqslant \pi$. [4]

10 Find the area of the segment shown. [3]

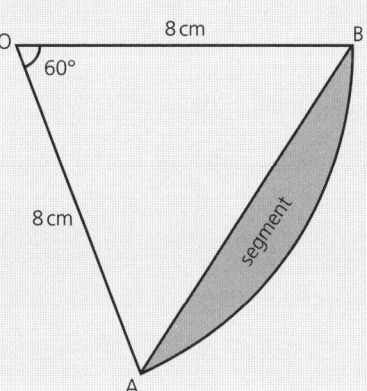

11 In the diagram, AC is an arc of a circle, centre O and radius 10 cm.

The line AB is perpendicular to OA and OCB is a straight line.

Angle AOC $= \frac{1}{3}\pi$ radians.

Find the area and perimeter of the shaded region, giving your answer in terms of π and $\sqrt{3}$. **[6]**

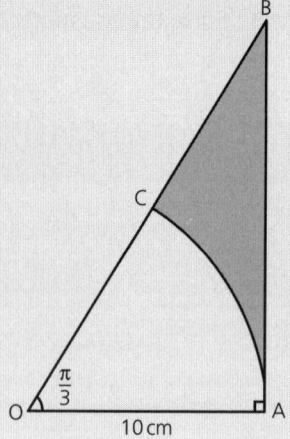

12 In the diagram, the circle has centre O and radius 6 cm.

The points D and F lie on the circle, and the arc length DF is 10 cm.

The tangents to the circle at D and F meet at the point E.

Calculate

(i) angle DOF in radians **[1]**

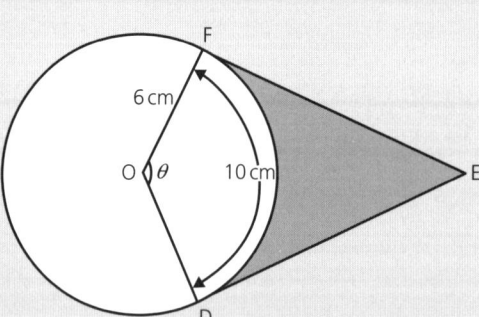

(ii) the length of DE [2]

(iii) the area of the shaded region. [3]

8 Vectors

Vector calculations

1 $\mathbf{a} = \begin{pmatrix} -3 \\ 1 \end{pmatrix}$, $\mathbf{b} = \begin{pmatrix} 5 \\ 0 \end{pmatrix}$ and $\mathbf{c} = \begin{pmatrix} 2 \\ -4 \end{pmatrix}$.

   (i) Draw and label each of the vectors on the grid below.

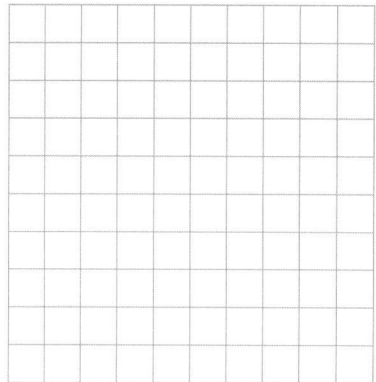

   (ii) Find each of the following vectors.

       Draw and label each of the vectors on the grid below.

$$2\mathbf{a} = \begin{pmatrix} \\ \end{pmatrix} \qquad -\mathbf{b} = \begin{pmatrix} \\ \end{pmatrix} \qquad \tfrac{1}{2}\mathbf{c} = \begin{pmatrix} \\ \end{pmatrix}$$

(iii) Find the following vector sums.

Draw and label diagrams on the grids below to represent the vector sums.

$\mathbf{a} + \mathbf{b} = \begin{pmatrix} \\ \end{pmatrix}$ 　　　　$\mathbf{b} - 2\mathbf{c} = \begin{pmatrix} \\ \end{pmatrix}$ 　　　　$\frac{1}{2}\mathbf{c} + 2\mathbf{a} = \begin{pmatrix} \\ \end{pmatrix}$

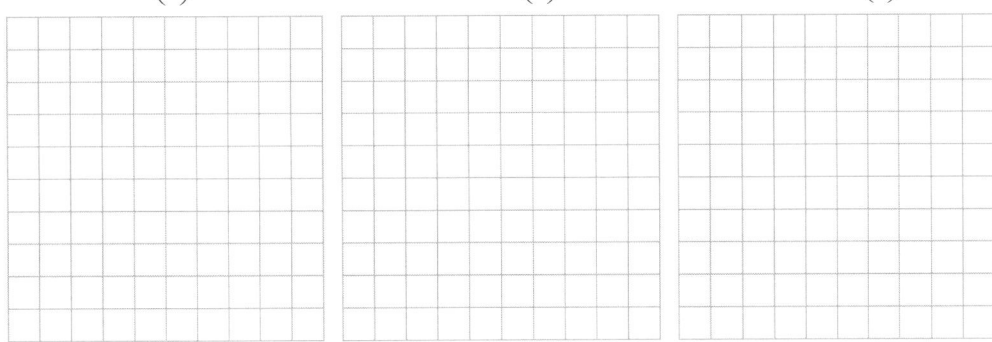

2 Given that $\mathbf{i} = \begin{pmatrix} 1 \\ 0 \end{pmatrix}$ and $\mathbf{j} = \begin{pmatrix} 0 \\ 1 \end{pmatrix}$ are the unit vectors parallel with the x and y axes

respectively, write the following vectors in terms of $\mathbf{i}$ and $\mathbf{j}$.

(i) $\begin{pmatrix} 4 \\ -5 \end{pmatrix} =$ 　　　　　　　　**(ii)** $\begin{pmatrix} -1 \\ 2 \end{pmatrix} =$

(iii) $\mathbf{g} =$

$\mathbf{h} =$

$\mathbf{g} + \mathbf{h}$

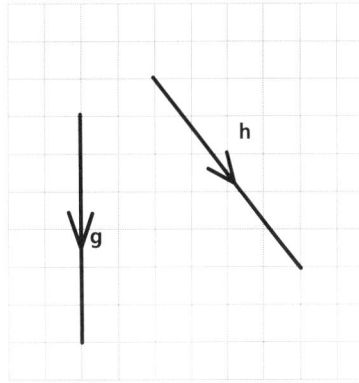

3 Find the length of the following vectors.

(i) $\begin{pmatrix} -3 \\ 2 \end{pmatrix}$ 　　　　　　　　**(ii)** $2\mathbf{i} - \mathbf{j}$

(iii) $\overrightarrow{AB}$ where A is (2, −1) and B is (−4, 0).

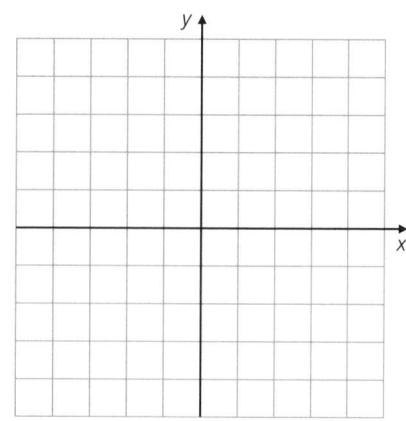

4 Use the axes below to plot the points A(−3, 2), B(4, −3) and C(0, 4).

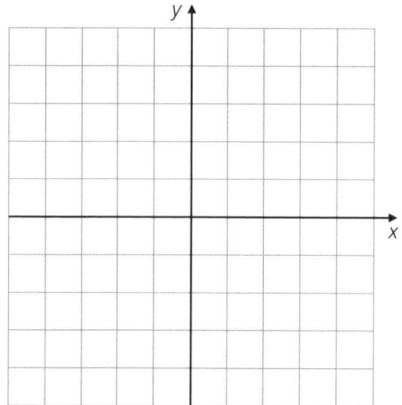

(i) Find the position vector $\overrightarrow{OA}$.

(ii) Find the displacement vector $\overrightarrow{BC}$.

(iii) Find the displacement vector $\overrightarrow{CA}$.

(iv) Find the co-ordinates of the point D such that $\overrightarrow{AD} = 3\overrightarrow{BC}$.

5 The diagram shows the vectors **m** and **n**.

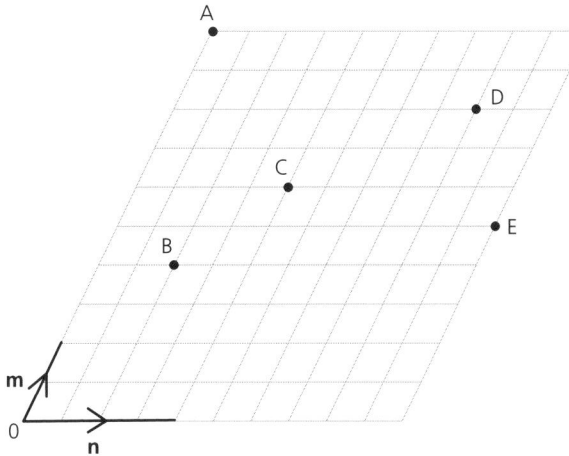

Find the following vectors in terms of **m** and **n**.

(i) $\vec{OB} =$

(ii) $\vec{OE} =$

(iii) $\vec{BD} =$

(iv) $\vec{EC} =$

6 Find the unit vector in the direction of the following vectors.

(i) $\begin{pmatrix} -4 \\ 3 \end{pmatrix}$

(ii) $\vec{EF}$ where E is $(-6, 2)$ and F is $(-2, 1)$.

7 If $\mathbf{d} = \begin{pmatrix} 1 \\ -2 \\ 3 \end{pmatrix}$ and $\mathbf{e} = \begin{pmatrix} 0 \\ 4 \\ -8 \end{pmatrix}$, find the vector for the following.

(i) $2\mathbf{d} + 3\mathbf{e} = \begin{pmatrix} \\ \\ \end{pmatrix}$ **(ii)** $3\mathbf{d} - \frac{1}{2}\mathbf{e} = \begin{pmatrix} \\ \\ \end{pmatrix}$

8 The position vector of the point A is $2\mathbf{i} - 3\mathbf{j} + \mathbf{k}$ and the position vector of the point B is $\mathbf{i} + 2\mathbf{j} - 3\mathbf{k}$.

Find the displacement vector $\overrightarrow{AB}$.

9 The diagram shows a gold bar.

The cross-section of the bar is in the shape of an isosceles trapezium.

Unit vectors $\mathbf{i}, \mathbf{j}$ and $\mathbf{k}$ are as shown.

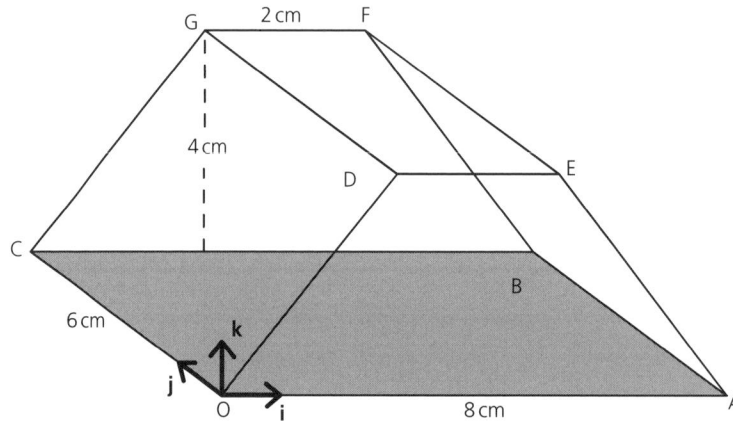

Find the following vectors in terms of $\mathbf{i}, \mathbf{j}$ and $\mathbf{k}$.

(i) $\overrightarrow{OD} =$ **(ii)** $\overrightarrow{OF} =$

(iii) $\overrightarrow{CE} =$ **(iv)** $\overrightarrow{BF} =$

(v) $\overrightarrow{GA} =$ **(vi)** $\overrightarrow{DF} =$

10 Find the length of $\overrightarrow{CD}$ where C is (2, 0, −3) and D is (−3, 2, 5).

11 Relative to the origin, the position vectors of the points P, Q and R are

$$\overrightarrow{OP} = \begin{pmatrix} 1 \\ -1 \\ 3 \end{pmatrix}, \ \overrightarrow{OQ} = \begin{pmatrix} 1 \\ 2 \\ -3 \end{pmatrix}, \ \overrightarrow{OR} = \begin{pmatrix} 1 \\ -4 \\ 9 \end{pmatrix}.$$

(i) Find the vector $\overrightarrow{PQ}$.   **(ii)** Find the vector $\overrightarrow{QR}$.

(iii) Hence explain the geometric relationship between $\overrightarrow{PQ}$ and $\overrightarrow{QR}$.

(iv) Another point, S, lies on the line that goes through R and P such that P is the mid-point of RS.

Find the co-ordinates of S.

12 Relative to an origin O, the points A and B have position vectors
$2\mathbf{i} - 2\mathbf{j} + 3\mathbf{k}$ and $4\mathbf{i} - 5\mathbf{k}$ respectively.

The point C is such that $\overrightarrow{OC} = 2\overrightarrow{OA}$.

The point D is such that $\overrightarrow{OD} = 3\overrightarrow{OB}$ and the point E is the mid-point of AB.

Find:

(i) $\overrightarrow{CD}$

(ii) $\overrightarrow{EC}$

13 The position vectors of the points E and F are $2\mathbf{i} - \mathbf{j} + 6\mathbf{k}$ and $3\mathbf{i} - 3\mathbf{j} + 4\mathbf{k}$ respectively.

The point G lies on the line EF such that the magnitude of EG is 18.

Find the co-ordinates of the point G.

14 Find the unit vector in the direction of

    (i) $-\mathbf{i} - 2\mathbf{j} + 2\mathbf{k}$

    (ii) $\overrightarrow{GH}$ where G is $(4, -1, 12)$ and H is $(9, -1, 0)$.

15 The points H and L have position vectors $\mathbf{h} = 2\mathbf{i} - \mathbf{j} + 6\mathbf{k}$ and $\mathbf{l} = 5\mathbf{i} - \mathbf{j} + 10\mathbf{k}$.

    The point M lies on the line HL such that $\overrightarrow{HM} = 3\overrightarrow{HL}$.

    Find the unit vector in the direction of $\mathbf{m}$, the position vector of the point M.

The angle between two vectors

1 Find the angle between the following vectors.

(i) $\begin{pmatrix} 3 \\ -4 \end{pmatrix}$ and $\begin{pmatrix} -1 \\ 2 \end{pmatrix}$

(ii) $3\mathbf{i} + 7\mathbf{j}$ and $\mathbf{i} - 6\mathbf{j}$

2 Find the angle AOB where A is the point $(2, -3, -1)$ and B is the point $(0, 5, -2)$.

3 The position vectors of the points C and D are given by

$$\overrightarrow{OC} = \begin{pmatrix} 1 \\ 3 \\ -2 \end{pmatrix} \text{ and } \overrightarrow{OD} = \begin{pmatrix} 6 \\ 0 \\ 5 \end{pmatrix}.$$

(i) Find the angle between the vectors $\overrightarrow{OC}$ and $\overrightarrow{OD}$.

(ii) Find the angle DCO, the angle between the vectors $\overrightarrow{OC}$ and $\overrightarrow{CD}$.

(iii) The vector $\overrightarrow{OF} = \begin{pmatrix} a \\ 3 \\ 2a \end{pmatrix}$ is perpendicular to $\overrightarrow{CD}$. Find the value of a.

4 The position vectors of the points E and F are $2\mathbf{i} + 3\mathbf{j} - 5\mathbf{k}$ and $3\mathbf{i} - 2\mathbf{j} - \mathbf{k}$ respectively.

Find the angle EOF.

5 Find the value of c such that the vectors $\mathbf{g} = 8\mathbf{i} + c\mathbf{j} - 2\mathbf{k}$ and $\mathbf{h} = c\mathbf{i} - 3\mathbf{j} - \mathbf{k}$ are perpendicular.

6 The diagram shows a right-angled triangular prism.

F is the mid-point of CE.

Unit vectors **i**, **j**, **k** are shown.

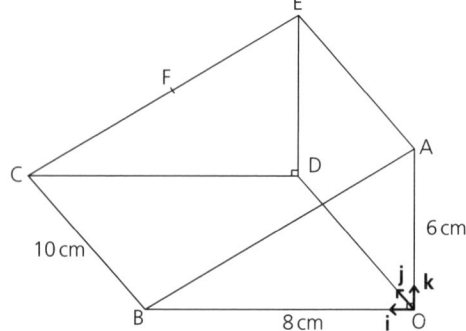

(i) Find the length of AB.

(ii) Express the vectors $\overrightarrow{AF}$ and $\overrightarrow{BF}$ in terms of **i**, **j** and **k**.

(iii) Hence find the angle between the vectors $\overrightarrow{AF}$ and $\overrightarrow{BF}$.

7 The diagram shows a bucket.

The base of the bucket is circular with centre O and radius 12 cm.

Unit vectors **i**, **j**, and **k** are shown parallel to OA, OB and OO′ respectively.

The vertical height of the bucket is 40 cm and the open circular top has a radius of 16 cm.

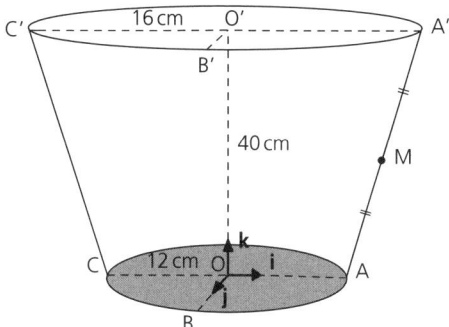

(i) Find, in terms of **i**, **j**, and **k**, the vectors $\overrightarrow{OM}$, $\overrightarrow{C'M}$, $\overrightarrow{B'M}$ and $\overrightarrow{CB'}$.

(ii) Hence find the angles OMC′ and CB′M.

Stretch and challenge

1 Find a non-zero vector which is perpendicular to both $-\mathbf{i} + \mathbf{j} + 3\mathbf{k}$ and $2\mathbf{i} + 5\mathbf{k}$.

2 The vectors $\mathbf{p}$ and $\mathbf{q}$ are given by $\begin{pmatrix} a \\ b \\ 3 \end{pmatrix}$ and $\begin{pmatrix} 4 \\ a \\ 3 \end{pmatrix}$ respectively.

Find the values of the constants a and b if $|\mathbf{p}| = |\mathbf{q}|$ and $\mathbf{p}$ and $\mathbf{q}$ are perpendicular.

3 Two planes A and B are flying on exactly the same straight-line path with plane B behind plane A.

Plane B is flying with a constant velocity of $(120\mathbf{i} - 60\mathbf{j} + 40\mathbf{k})$ m/s.

Plane A is moving away from Plane B at a constant rate of 35 m/s.

Find the vector representing the velocity of Plane A.

■ *Exam focus*

1 The position vectors of the points P and Q are $\overrightarrow{OP} = \begin{pmatrix} 2 \\ -1 \\ 4 \end{pmatrix}$ and $\overrightarrow{OQ} = \begin{pmatrix} 10 \\ 1 \\ 12 \end{pmatrix}$.

The point R lies on the line PQ such that $\overrightarrow{PR} = \frac{1}{2}\overrightarrow{PQ}$.

Find the unit vector in the direction of $\overrightarrow{OR}$.   **[4]**

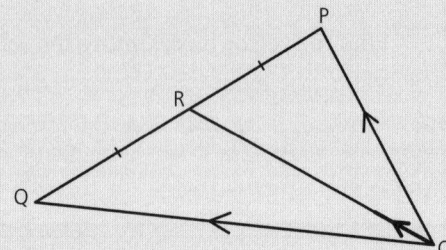

2 The position vectors of the points M and N are given by

$$\overrightarrow{OM} = \begin{pmatrix} 2 \\ 3 \\ -1 \end{pmatrix} \text{ and } \overrightarrow{ON} = \begin{pmatrix} p \\ -2 \\ 3p \end{pmatrix}.$$

Find the value of p if $\overrightarrow{OM}$ is perpendicular to $\overrightarrow{ON}$.   **[3]**

3 The diagram shows a gold bar.

The cross-section of the bar is in the shape of an isosceles trapezium.

The point M is the mid-point of AB.

Unit vectors **i**, **j** and **k** are as shown.

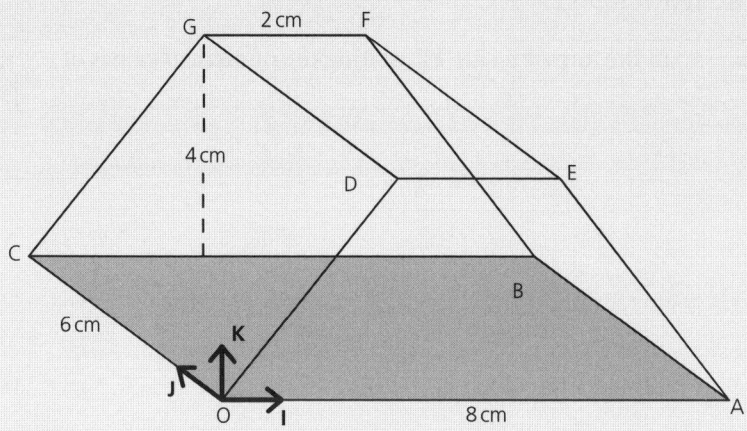

Evaluate $\overrightarrow{DF} \cdot \overrightarrow{DM}$ and hence find angle FDM. [5]

PAST EXAMINATION QUESTIONS

1 Algebra

1 The equation of a curve is $y = 8x - x^2$.

   (i) Express $8x - x^2$ in the form $a - (x + b)^2$, stating the numerical values of
      a and b.   **[3]**

....

....

....

....

   (ii) Find the set of values of x for which $y \geqslant -20$.   **[3]**

....

....

....

....

....

                                                                 [Total: 6]

Cambridge International AS & A Level Mathematics, 9709/01 June 2003 Q 11 i, iii

2 The curve y is defined by $y = x^2 - 3x$ for $x \in \mathbb{R}$.

   (i) Find the set of values of x for which $y > 4$.   **[3]**

....

....

....

....

....

....

(ii) Express y in the form $(x - a)^2 - b$, stating the values of a and b. [2]

....

....

....

[Total: 5]

Adapted from Cambridge International AS & A Level Mathematics, 9709/01 November 2006
Q 10 i, ii

2 Co-ordinate geometry

1 The diagram shows the rectangle *ABCD*, where *A* is (3, 2) and *B* is (1, 6).

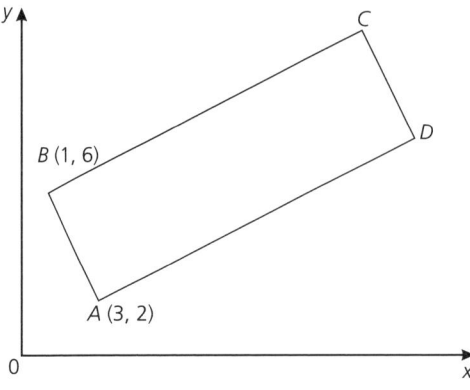

(i) Find the equation of *BC*. [4]

....

....

....

Given that the equation of *AC* is $y = x - 1$, find

(ii) the co-ordinates of *C*, [2]

....

....

....

....

....

....

(iii) the perimeter of the rectangle *ABCD*. [3]

....

....

....

[Total: 9]

Cambridge International AS & A Level Mathematics, 9709/01 November 2002 Q 9

2 The diagram shows a trapezium *ABCD* in which *BC* is parallel to *AD* and angle *BCD* = 90°. The co-ordinates of *A*, *B* and *D* are (2, 0), (4, 6) and (12, 5) respectively.

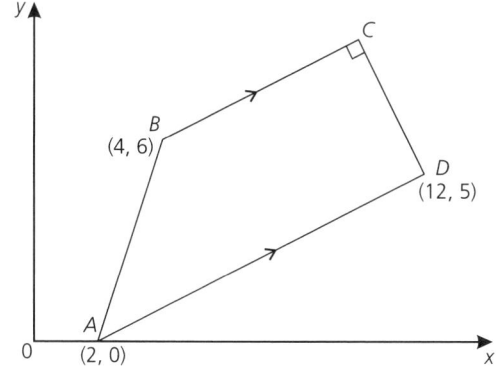

(i) Find the equations of *BC* and *CD*.   [5]

....

....

....

....

....

....

(ii) Calculate the co-ordinates of *C*.   [2]

....

....

....

....

....

....

[Total: 7]

Cambridge International AS & A Level Mathematics, 9709/01 November 2003 Q 5

3 The equation of a curve is $xy = 12$ and the equation of a line l is $2x + y = k$, where k is a constant.

(i) In the case where $k = 11$, find the co-ordinates of the points of intersection of l and the curve. **[3]**

....

....

....

....

....

....

....

....

....

....

(ii) Find the set of values of k for which l does not intersect the curve. **[4]**

....

....

....

....

....

....

....

....

....

....

[Total: 7]

Cambridge International AS & A Level Mathematics, 9709/01 November 2005 Q 9 i, ii

4 The three points $A(1, 3)$, $B(13, 11)$ and $C(6, 15)$ are shown in the diagram. The perpendicular from C to AB meets AB at the point D.

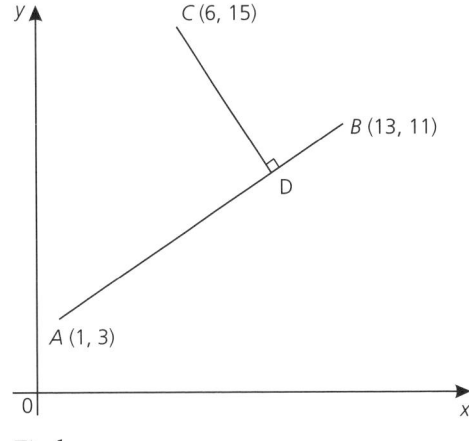

Find

(i) the equation of CD, [3]

....

....

....

....

....

....

(ii) the co-ordinates of D. [4]

....

....

....

....

....

....

[Total: 7]

Cambridge International AS & A Level Mathematics, 9709/01 November 2006 Q 5

5 Find the set of values of m for which the line $y = mx + 4$ intersects the curve
$y = 3x^2 - 4x + 7$ at two distinct points. [5]

....

....

....

....

....

....

Cambridge International AS & A Level Mathematics, 9709/13 May/June 2011 Q 2

6 The diagram shows a quadrilateral $ABCD$ in which the point A is $(-1, -1)$,
the point B is $(3, 6)$ and the point C is $(9, 4)$. The diagonals AC and BD intersect
at M. Angle $BMA = 90°$ and $BM = MD$.

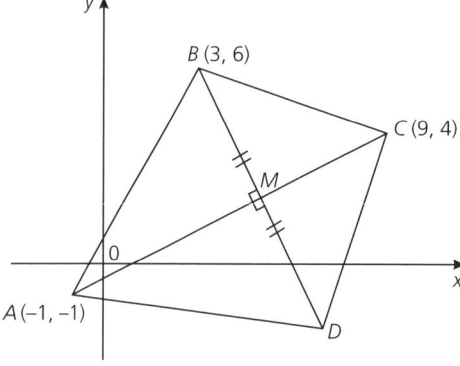

Calculate:

(i) the co-ordinates of M and D. [7]

....

....

....

....

....

....

(ii) the ratio $AM : MC$. [2]

....

....

....

....

....

....

[Total: 9]

Cambridge International AS & A Level Mathematics, 9709/12 November 2011 Q 9

7 The equation of a curve is $y^2 + 2x = 13$ and the equation of a line is $2y + x = k$, where k is a constant.

(i) In the case where $k = 8$, find the co-ordinates of the points of intersection of the line and the curve. [4]

....

....

....

....

....

....

....

....

(ii) Find the value of k for which the line is a tangent to the curve. [3]

....

....

....

....

....

....

....

....

....

....

[Total: 7]

Cambridge International AS & A Level Mathematics, 9709/12 November 2011 Q 4

3 Sequences and series

1 A geometric progression, for which the common ratio is positive, has a second term of 18 and a fourth term of 8. Find

(i) the first term and the common ratio of the progression, [3]

....

....

....

....

....

....

....

(ii) the sum to infinity of the progression. [2]

....

....

....

....

[Total: 5]

Cambridge International AS & A Level Mathematics, 9709/01 November 2002 Q 2

2 (i) A debt of \$3726 is repaid by weekly payments which are in arithmetic progression. The first payment is \$60 and the debt is fully repaid after 48 weeks. Find the third payment. **[3]**

....

....

....

....

....

....

....

....

(ii) Find the sum to infinity of the geometric progression whose first term is 6 and whose second term is 4. **[3]**

....

....

....

....

....

....

....

....

[Total: 6]

Cambridge International AS & A Level Mathematics, 9709/01 November 2003 Q 3

3 Find the coefficient of x in the expansion of $\left(3x - \dfrac{2}{x}\right)^5$. [4]

....

....

....

....

....

....

....

Cambridge International AS & A Level Mathematics, 9709/01 November 2004 Q 1

4 Find

(i) the sum of the first ten terms of the geometric progression 81, 54, 36, ... , [3]

....

....

....

....

(ii) the sum of all the terms in the arithmetic progression 180, 175, 170, ... , 25. [3]

....

....

....

....

....

....

[Total: 6]

Cambridge International AS & A Level Mathematics, 9709/01 November 2004 Q 2

5 A small trading company made a profit of \$250 000 in the year 2000. The company considered two different plans, plan *A* and plan *B*, for increasing its profits.
Under plan *A*, the annual profit would increase each year by 5% of its value in the preceding year.

Find, for plan *A*,

(i) the profit for the year 2008,   [3]

....

....

....

....

(ii) the total profit for the 10 years 2000 to 2009 inclusive.   [2]

....

....

....

Under plan *B*, the annual profit would increase each year by a constant amount \$*D*.

(iii) Find the value of *D* for which the total profit for the 10 years 2000 to 2009 inclusive would be the same for both plans.   [3]

....

....

....

....

....

....

....

....

[Total: 8]

Cambridge International AS & A Level Mathematics, 9709/01 November 2005 Q 6

6 (i) Find the sum of all the integers between 100 and 400 that are divisible by 7. [4]

....

....

....

....

....

(ii) The first three terms in a geometric progression are 144, x and 64 respectively, where x is positive.

Find

(a) the value of x, [3]

....

....

....

....

(b) the sum to infinity of the progression. [2]

....

....

....

....

[Total: 9]

Cambridge International AS & A Level Mathematics, 9709/01 November 2006 Q 6

7 (i) Find the first three terms in the expansion of $(2 + u)^5$ in ascending powers of u. [3]

....

....

....

(ii) Use the substitution $u = x + x^2$ in your answer to part **(i)** to find the coefficient of x^2 in the expansion of $\left(2 + x + x^2\right)^5$. [2]

....

....

....

....

....

....

....

[Total: 5]

Cambridge International AS & A Level Mathematics, 9709/01 November 2007 Q 3

8 (i) Find the first three terms in the expansion, in ascending powers of x, of $\left(2 + x^2\right)^5$. [3]

....

....

....

(ii) Hence find the coefficient of x^4 in the expansion of $\left(1 + x^2\right)^2\left(2 + x^2\right)^5$. [3]

....

....

....

....

....

....

....

[Total: 6]

Cambridge International AS & A Level Mathematics, 9709/01 June 2008 Q 3

9 The coefficient of x^3 in the expansion of $(a + x)^5 + (1 - 2x)^6$, where a is positive, is 90. Find the value of a. [5]

....

....

....

....

....

....

....

....

....

....

Cambridge International AS & A Level Mathematics, 9709/13 June 2011 Q 1

10 **(i)** An arithmetic progression contains 25 terms and the first term is −15. The sum of all the terms in the progression is 525. Calculate

(a) the common difference of the progression, [2]

....

....

....

....

....

....

(b) the last term in the progression, [2]

....

....

(c) the sum of all the positive terms in the progression. [2]

....

....

....

....

....

....

(ii) A college agrees a sponsorship deal in which grants will be received each year for sports equipment. This grant will be $4000 in 2012 and will increase by 5% each year. Calculate

(a) the value of the grant in 2022, [2]

....

....

(b) the total amount the college will receive in the years 2012 to 2022 inclusive. [2]

....

....

....

[Total:10]

Cambridge International AS & A Level Mathematics, 9709/12 November 2011 Q 10

4 Functions

1 The functions f and g are defined by

$$f : x \mapsto 3x + 2, \qquad x \in \mathbb{R}$$

$$g : x \mapsto \frac{6}{2x + 3}, \qquad x \in \mathbb{R}, \qquad x \neq -1.5.$$

(i) Find the value of x for which $fg(x) = 3$. [3]

....

....

....

....

....

(ii) Sketch, in a single diagram, the graphs of $y = f(x)$ and $y = f^{-1}(x)$, making clear the relationship between the two graphs. [3]

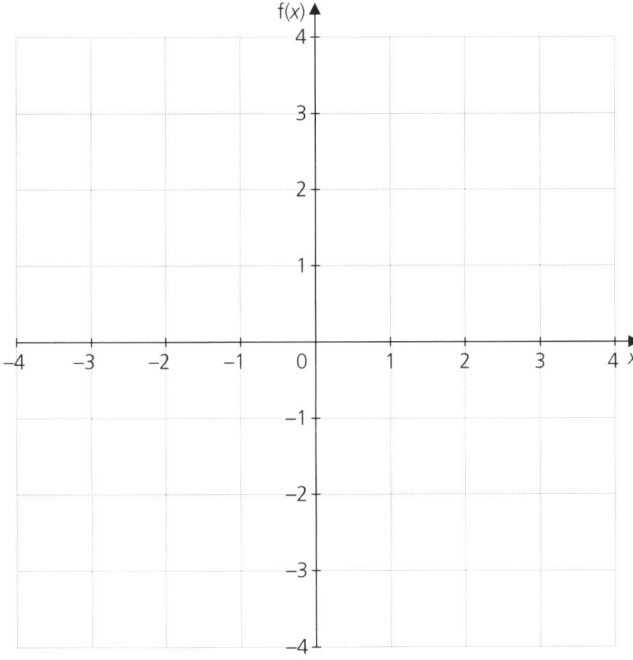

(iii) Express each of $f^{-1}(x)$ and $g^{-1}(x)$ in terms of x, and solve the equation $f^{-1}(x) = g^{-1}(x)$. **[5]**

....

....

....

....

....

....

....

....

....

[Total: 11]

Cambridge International AS & A Level Mathematics, 9709/01 June 2002 Q 10

2 The function f is defined by $f : x \mapsto ax + b$, for $x \in \mathbb{R}$, where a and b are constants. It is given that $f(2) = 1$ and $f(5) = 7$.

(i) Find the values of a and b. **[2]**

....

....

....

....

....

(ii) Solve the equation $ff(x) = 0$. **[3]**

....

....

....

....

....

[Total: 5]

Cambridge International AS & A Level Mathematics, 9709/01 June 2003 Q 5

3 The function g is defined by $g : x \mapsto 8x - x^2$, for $x \geqslant 4$.

(i) State the domain and range of g^{-1}. [2]

....

....

....

....

(ii) Find an expression, in terms of x, for $g^{-1}(x)$. [3]

....

....

....

....

[Total: 5]

Cambridge International AS & A Level Mathematics, 9709/01 June 2003 Q 11 iv, v

4 The functions f and g are defined as follows:

$$f : x \mapsto x^2 - 2x, \quad x \in \mathbb{R}$$
$$g : x \mapsto 2x + 3, \quad x \in \mathbb{R}$$

(i) Find the set of values of x for which $f(x) > 15$. [3]

....

....

....

....

....

....

....

(ii) Find the range of f and state, with a reason, whether f has an inverse. **[4]**

....

....

....

....

....

(iii) Show that the equation $gf(x) = 0$ has no real solutions. **[3]**

....

....

....

....

....

(iv) Sketch in a single diagram the graphs of $y = g(x)$ and $y = g^{-1}(x)$, making clear the relationship between the graphs. **[2]**

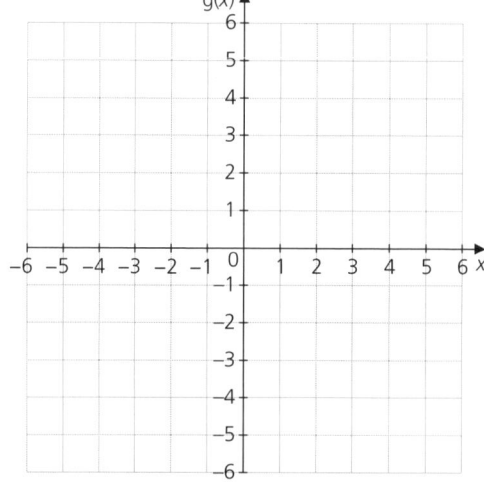

[Total: 12]

Cambridge International AS & A Level Mathematics, 9709/01 June 2004 Q 10

5 A function f is defined by $f : x \mapsto 3 - 2\sin x$, for $0° \leqslant x \leqslant 360°$.

(i) Find the range of f. [2]

....

....

(ii) Sketch the graph of $y = f(x)$. [2]

A function g is defined by $g : x \mapsto 3 - 2\sin x$, for $0° \leqslant x \leqslant A°$, where A is a constant.

(iii) State the largest value of A for which g has an inverse. [1]

....

(iv) When A has this value, obtain an expression, in terms of x, for $g^{-1}(x)$. [2]

....

....

....

....

....

[Total: 7]

Cambridge International AS & A Level Mathematics, 9709/01 June 2004 Q 7

6 The function f is defined by $f(x) = x^2 - 3x = \left(x - \frac{3}{2}\right)^2 - \frac{9}{4}$ for $x \in \mathbb{R}$.

(i) Write down the range of f. [1]

....

....

(ii) State, with a reason, whether f has an inverse. [1]

....

....

The function g is defined by $g : x \mapsto x - 3\sqrt{x}$ for $x \geqslant 0$.

(iii) Solve the equation $g(x) = 10$. [3]

....

....

....

....

....

[Total: 5]

Adapted from Cambridge International AS & A Level Mathematics, 9709/01 November 2006
Q 10 iii, iv, v

7 The diagram shows the graph of $y = f(x)$, where $f : x \mapsto \dfrac{6}{2x+3}$ for $x \geq 0$.

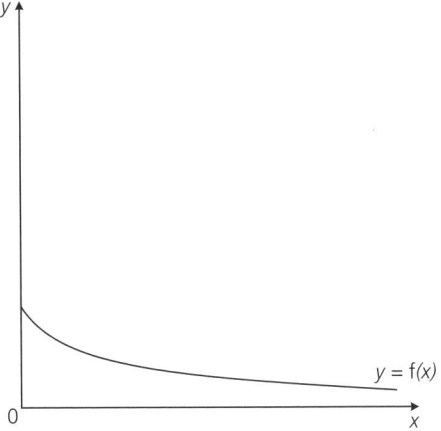

(i) Find an expression, in terms of x, for $f^{-1}(x)$ and find the domain of f^{-1}.　　**[4]**

....

....

....

....

....

....

(ii) On the diagram above sketch the graph of $y = f^{-1}(x)$, making clear the relationship between the graphs.　　**[2]**

The function g is defined by $g : x \mapsto \dfrac{1}{2}x$ for $x \geq 0$.

(iii) Solve the equation $fg(x) = \dfrac{3}{2}$.　　**[3]**

....

....

....

....

....

....

[Total: 9]

Adapted from Cambridge International AS & A Level Mathematics, 9709/01 June 2007
Q 11 ii, iii, iv

8 The function f is such that $f(x) = (3x + 2)^3 - 5$ for $x \geqslant 0$.

Obtain an expression for $f^{-1}(x)$ and state the domain of f^{-1}. [4]

Cambridge International AS & A Level Mathematics, 9709/01 June 2008 Q 6

9 The diagram opposite shows the function f defined for $0 \leqslant x \leqslant 6$ by

$x \mapsto \dfrac{1}{2}x^2$ for $0 \leqslant x \leqslant 2$,

$x \mapsto \dfrac{1}{2}x + 1$ for $2 < x \leqslant 6$.

(i) State the range of f. [1]

(ii) On the diagram sketch the graph of $y = \mathrm{f}^{-1}(x)$. [2]

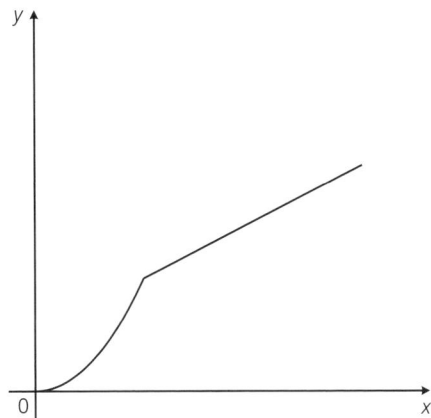

(iii) Obtain expressions to define $\mathrm{f}^{-1}(x)$, giving the set of values of x for which each expression is valid. [4]

....

....

....

....

....

[Total: 7]

Adapted from Cambridge International AS & A Level Mathematics, 9709/13 November 2010 Q 7

10 The functions f and g are defined for $x \in \mathbb{R}$ by

$\mathrm{f} : x \mapsto 3x + a,$

$\mathrm{g} : x \mapsto b - 2x,$

where a and b are constants. Given that $\mathrm{ff}(2) = 10$ and $\mathrm{g}^{-1}(2) = 3$, find

(i) the values of a and b, [4]

....

....

....

....

....

....

....

....

(ii) an expression for fg(x). [2]

....

....

....

....

....

....

[Total: 6]

Cambridge International AS & A Level Mathematics, 9709/12 November 2011 Q 2

5 Differentiation

1 A hollow circular cylinder, open at one end, is constructed of thin sheet metal. The total external surface area of the cylinder is $192\pi \, \text{cm}^2$. The cylinder has a radius of r cm and a height of h cm.

(i) Express h in terms of r and show that the volume, $V \, \text{cm}^3$, of the cylinder is given by

$$V = \frac{1}{2}\pi(192r - r^3)$$ [4]

Given that r can vary,

(ii) find the value of r for which V has a stationary value,   [3]

(iii) find this stationary value and determine whether it is a maximum or a minimum. [3]

....

....

....

....

....

....

[Total: 10]

Cambridge International AS & A Level Mathematics, 9709/01 June 2002 Q 8

2 The diagram shows a glass window consisting of a rectangle of height h m and width $2r$ m and a semicircle of radius r m. The perimeter of the window is 8 m.

(i) Express h in terms of r. [2]

....

....

....

....

....

(ii) Show that the area of the window, $A \, \text{m}^2$, is given by

$$A = 8r - 2r^2 - \frac{1}{2}\pi r^2.$$ [2]

....

....

....

....

....

....

Given that r can vary,

(iii) find the value of r for which A has a stationary value, [4]

....

....

....

....

....

....

....

....

(iv) determine whether this stationary value is a maximum or a minimum. [2]

....

....

....

....

[Total: 10]

Cambridge International AS & A Level Mathematics, 9709/01 June 2004 Q8

3 The equation of a curve is $y = x^2 - 4x + 7$. Find the coordinates of the point Q on the curve at which the tangent is parallel to the line $y + 3x = 9$. **[3]**

Adapted from Cambridge International AS & A Level Mathematics, 9709/01 November 2004 Q 5 ii

4 Find the gradient of the curve $y = \dfrac{12}{x^2 - 4x}$ at the point where $x = 3$. **[4]**

Cambridge International AS & A Level Mathematics, 9709/01 June 2005 Q 2

5 The equation of a curve is $xy = 12$ and the equation of a line l is $2x + y = k$, where k is a constant. In the case where $k = 10$, one of the points of intersection is $P(2, 6)$. Find the angle, in degrees correct to 1 decimal place, between l and the tangent to the curve at P. [4]

Cambridge International AS & A Level Mathematics, 9709/01 November 2005 Q 9 iii

6 The diagram shows the graph of $y = f(x)$, where $f : x \mapsto \dfrac{6}{2x + 3}$ for $x \geqslant 0$.

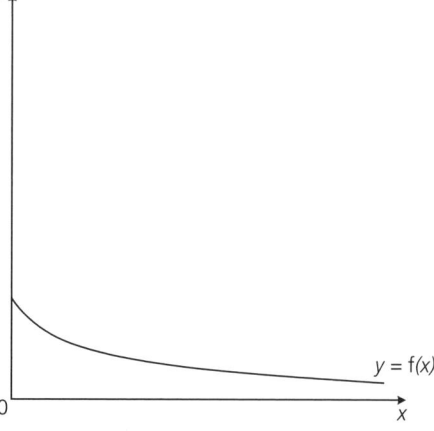

Find an expression, in terms of x, for $f'(x)$ and explain how your answer shows that f is a decreasing function. **[3]**

....

....

....

....

....

....

Cambridge International AS & A Level Mathematics, 9709/01 June 2007 Q 11 i

7 The function f is such that $f(x) = (3x + 2)^3 - 5$ for $x \geqslant 0$.

Obtain an expression for $f'(x)$ and hence explain why f is an increasing function. **[3]**

....

....

....

....

....

....

....

Cambridge International AS & A Level Mathematics, 9709/01 June 2008 Q 6 i

8 The equation of a curve is $y = \dfrac{12}{x^2 + 3}$.

   (i) Obtain an expression for $\dfrac{dy}{dx}$.   **[2]**

   ....

   ....

   ....

   (ii) Find the equation of the normal to the curve at the point $P(1, 3)$.   **[3]**

   ....

   ....

   ....

   ....

   (iii) A point is moving along the curve in such a way that the x-coordinate is increasing at a constant rate of 0.012 units per second. Find the rate of change of the y-coordinate as the point passes through P.   **[2]**

   ....

   ....

[Total: 7]

Cambridge International AS & A Level Mathematics, 9709/11 November 2009 Q 7

9 A curve is such that $\dfrac{dy}{dx} = 3x^{\frac{1}{2}} - 6$ and the point $(9, 2)$ lies on the curve.

Find the x-coordinate of the stationary point on the curve and determine the nature of the stationary point.   **[3]**

....

....

....

....

....

....

....

....

Cambridge International AS & A Level Mathematics, 9709/11 June 2010 Q 6 ii

6 Integration

These questions may contain both differentiation and integration parts.

1 The gradient at any point (x, y) on a curve is $\sqrt{(1 + 2x)}$. The curve passes through the point (4, 11). Find

(i) the equation of the curve, [4]

....

....

....

....

....

....

....

(ii) the point at which the curve intersects the y-axis. [2]

....

....

....

....

[Total: 6]

Cambridge International AS & A Level Mathematics, 9709/01 November 2002 Q 4

2 The diagram shows the points $A(1, 2)$ and $B(4, 4)$ on the curve $y = 2\sqrt{x}$. The line BC is the normal to the curve at B, and C lies on the x-axis. Lines AD and BE are perpendicular to the x-axis.

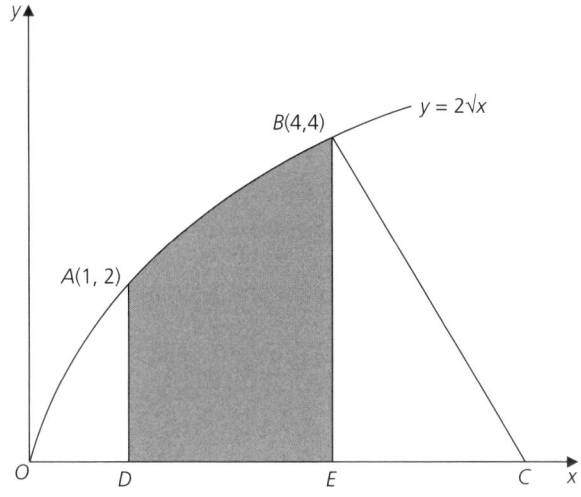

(i) Find the equation of the normal BC. [4]

....

....

....

....

....

....

(ii) Find the area of the shaded region. [4]

....

....

....

....

....

....

[Total: 8]

Cambridge International AS & A Level Mathematics, 9709/01 November 2002 Q 10

3 The equation of a curve is $y = \sqrt{(5x + 4)}$.

(i) Calculate the gradient of the curve at the point where $x = 1$. **[3]**

(ii) A point with co-ordinates (x, y) moves along the curve in such a way that the rate of increase of x has the constant value 0.03 units per second. Find the rate of increase of y at the instant when $x = 1$. **[2]**

(iii) Find the area enclosed by the curve, the x-axis, the y-axis and the line $x = 1$. **[5]**

[Total: 10]

Cambridge International AS & A Level Mathematics, 9709/01 June 2003 Q 10

4 A curve is such that $\dfrac{dy}{dx} = 3x^2 - 4x + 1$. The curve passes through the point (1, 5).

(i) Find the equation of the curve. [3]

....

....

....

....

....

....

(ii) Find the set of values of x for which the gradient of the curve is positive. [3]

....

....

....

....

....

....

[Total: 6]

Cambridge International AS & A Level Mathematics, 9709/01 November 2003 Q 4

5 The diagram shows points $A(0, 4)$ and $B(2, 1)$ on the curve $y = \dfrac{8}{3x + 2}$. The tangent to the curve at B crosses the x axis at C. The point D has co-ordinates (2, 0).

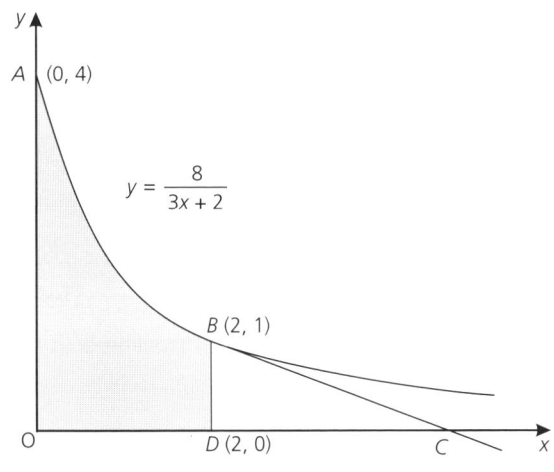

(i) Find the equation of the tangent to the curve at B and hence show that the area of triangle BDC is $\frac{4}{3}$. **[6]**

....

....

....

....

....

....

....

....

....

....

....

(ii) Show that the volume of the solid formed when the shaded region $ODBA$ is rotated completely about the x axis is 8π. **[5]**

....

....

....

....

....

....

....

....

....

....

[Total: 11]

Cambridge International AS & A Level Mathematics, 9709/01 November 2003 Q 9

6 A curve is such that $\dfrac{dy}{dx} = \dfrac{6}{\sqrt{(4x-3)}}$ and $P(3, 3)$ is a point on the curve.

   (i) Find the equation of the normal to the curve at P, giving your answer in the form $ax + by = c$.   **[3]**

   (ii) Find the equation of the curve.   **[4]**

                                                      [Total: 7]

Cambridge International AS & A Level Mathematics, 9709/01 November 2004 Q 7

7 The diagram shows the curve $y = x^3 - 3x^2 - 9x + k$, where k is a constant. The curve has a minimum point on the x-axis.

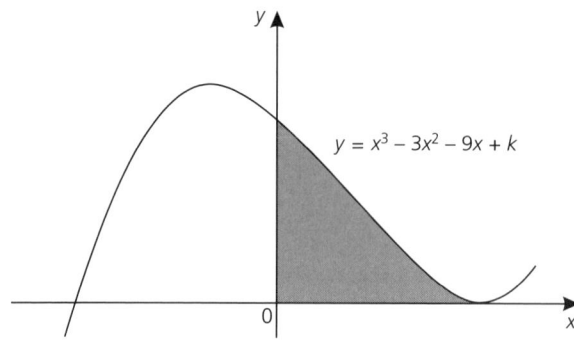

$y = x^3 - 3x^2 - 9x + k$

(i) Find the value of k. [4]

....

....

....

....

....

....

....

....

....

....

(ii) Find the co-ordinates of the maximum point of the curve. [1]

....

....

(iii) State the set of values of x for which $x^3 - 3x^2 - 9x + k$ is a decreasing function of x. [1]

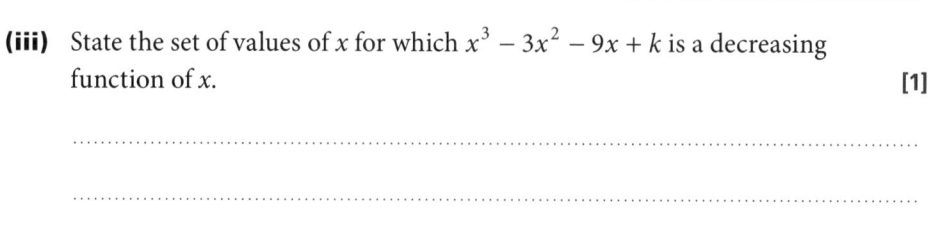

(iv) Find the area of the shaded region. [4]

....

....

....

....

....

[Total: 10]

Cambridge International AS & A Level Mathematics, 9709/01 June 2006 Q 10

8 The equation of a curve is $y = \dfrac{6}{5 - 2x}$.

(i) Calculate the gradient of the curve at the point where $x = 1$. [3]

....

....

....

....

(ii) A point with co-ordinates (x, y) moves along the curve in such a way that the rate of increase of y has a constant value of 0.02 units per second.
Find the rate of increase of x when $x = 1$. [2]

....

....

....

....

(iii) The region between the curve, the x-axis and the lines $x = 0$ and $x = 1$ is rotated through 360° about the x-axis. Show that the volume obtained is $\dfrac{12}{5}\pi$. [5]

....

....

....

....

....

....

....

[Total: 10]

Cambridge International AS & A Level Mathematics, 9709/01 November 2006 Q 8

9 The diagram shows the curve $y = \sqrt{(3x + 1)}$ and the points $P(0, 1)$ and $Q(1, 2)$ on the curve. The shaded region is bounded by the curve, the y-axis and the line $y = 2$.

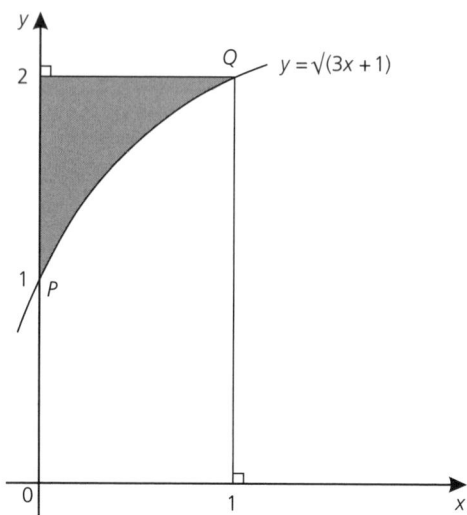

(i) Find the area of the shaded region. [4]

....

....

....

....

....

....

(ii) Find the volume obtained when the shaded region is rotated through 360° about the *x*-axis. [4]

....

....

....

....

....

....

....

....

....

....

....

Tangents are drawn to the curve at the points *P* and *Q*.

(iii) Find the acute angle, in degrees correct to 1 decimal place, between the two tangents. [4]

....

....

....

....

....

....

....

....

....

....

....

....

....

[Total: 12]

Cambridge International AS & A Level Mathematics, 9709/01 November 2008 Q 9

10 A curve is such that $\dfrac{\mathrm{d}y}{\mathrm{d}x} = k - 2x$, where k is a constant.

(i) Given that the tangents to the curve at the points where $x = 2$ and $x = 3$ are perpendicular, find the value of k. **[4]**

....

....

....

....

....

....

....

....

....

....

(ii) Given also that the curve passes through the point (4, 9), find the equation of the curve. **[3]**

....

....

....

....

....

....

[Total: 7]

Cambridge International AS & A Level Mathematics, 9709/11 November 2009 Q 6

11 The diagram shows parts of the curves $y = 9 - x^3$ and $y = \dfrac{8}{x^3}$ and their points of intersection P and Q. The x-coordinates of P and Q are a and b respectively.

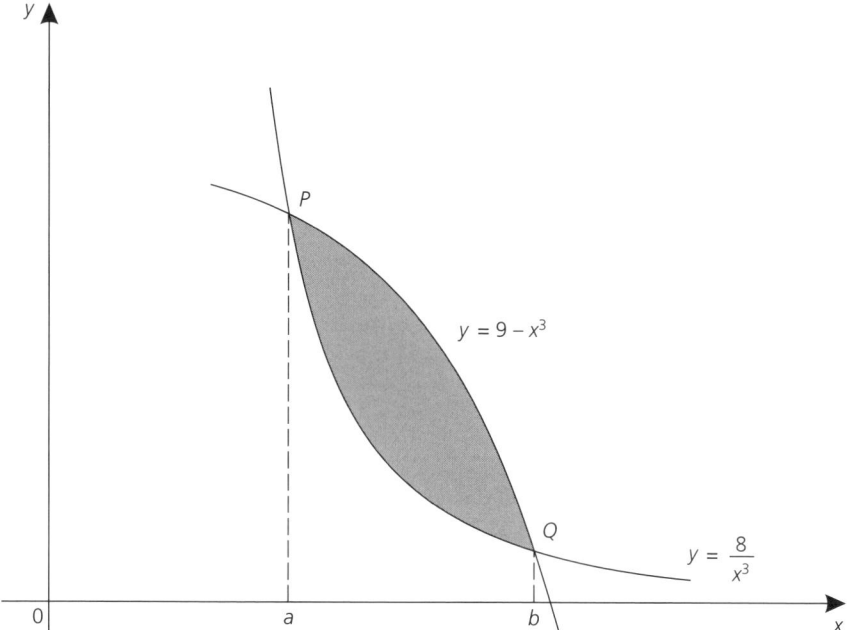

(i) Show that $x = a$ and $x = b$ are roots of the equation $x^6 - 9x^3 + 8 = 0$. Solve this equation and hence state the value of a and the value of b. **[4]**

....

....

....

....

....

....

....

....

....

....

(ii) Find the area of the shaded region between the two curves. [5]

....

....

....

....

....

....

....

....

....

....

(iii) The tangents to the two curves at $x = c$ (where $a < c < b$) are parallel to each other. Find the value of c. [4]

....

....

....

....

....

....

....

....

[Total: 13]

Cambridge International AS & A Level Mathematics, 9709/13 November 2010 Q 11

7 Trigonometry

1 The function f, where $f(x) = a \sin x + b$, is defined for the domain $0 \leqslant x \leqslant 2\pi$.
Given that $f\left(\frac{1}{2}\pi\right) = 2$ and that $f\left(\frac{3}{2}\pi\right) = -8$,

(i) find the values of a and b, [3]

....

....

....

....

....

....

....

(ii) find the values of x for which $f(x) = 0$, giving your answers in radians correct to 2 decimal places, [2]

....

....

....

(iii) sketch the graph of $y = f(x)$. [2]

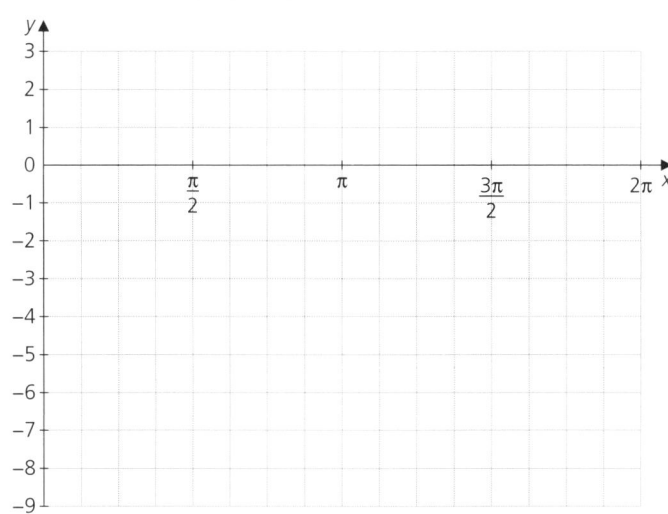

[Total: 7]

Cambridge International AS & A Level Mathematics, 9709/01 June 2002 Q 6

2 The diagram shows the circular cross-section of a uniform cylindrical log with centre O and radius 20 cm. The points A, X and B lie on the circumference of the cross-section and $AB = 32$ cm.

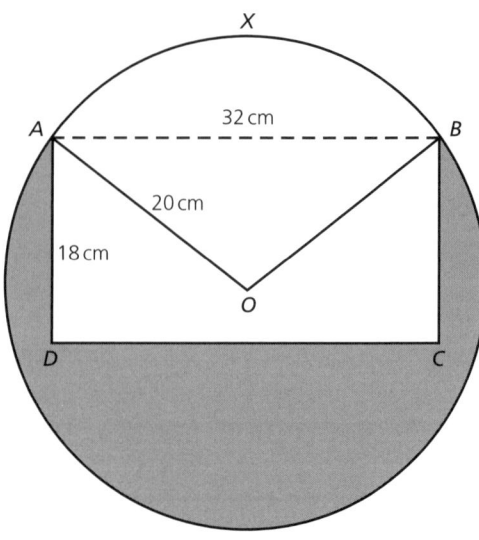

(i) Show that angle $AOB = 1.855$ radians, correct to 3 decimal places. **[2]**

....

....

....

(ii) Find the area of the sector $AXBO$. **[2]**

....

....

....

The section $AXBCD$, where $ABCD$ is a rectangle with $AD = 18$ cm, is removed.

(iii) Find the area of the new cross-section (shown shaded in the diagram). **[3]**

....

....

....

....

....

[Total: 7]

Cambridge International AS & A Level Mathematics, 9709/01 June 2002 Q 7

3 (i) Show that the equation $3\tan\theta = 2\cos\theta$ can be expressed as

$2\sin^2\theta + 3\sin\theta - 2 = 0$ [3]

....

....

....

....

....

....

....

(ii) Hence solve $3\tan\theta = 2\cos\theta$, for $0° \leqslant \theta \leqslant 360°$. [3]

....

....

....

....

....

....

....

....

[Total: 7]

Cambridge International AS & A Level Mathematics, 9709/01 November 2002 Q 5

4 In the diagram, triangle ABC is right-angled and D is the mid-point of BC. Angle $DAC = 30°$ and angle $BAD = x°$. Denoting the length of AD by l,

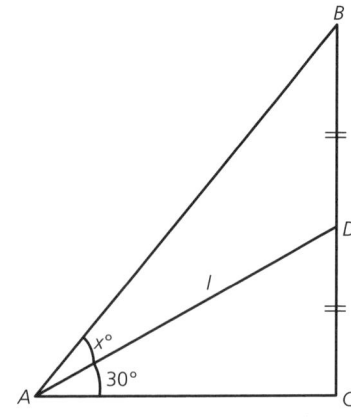

(i) express each of AC and BC exactly in terms of l, and show that $AB = \frac{1}{2}l\sqrt{7}$, **[4]**

....

....

....

....

....

....

....

....

(ii) show that $x = \tan^{-1}\dfrac{2}{\sqrt{3}} - 30$. **[2]**

....

....

....

....

....

....

....

....

[Total: 6]

Cambridge International AS & A Level Mathematics, 9709/01 November 2002 Q 6

5 The diagram shows a semicircle ABC with centre O and radius 8 cm.
Angle $AOB = \theta$ radians.

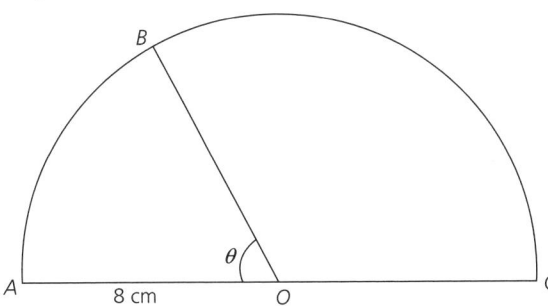

(i) In the case where $\theta = 1$, calculate the area of the sector BOC. [3]

....

....

....

....

(ii) Find the value of θ for which the perimeter of sector AOB is one half of the perimeter of sector BOC. [3]

....

....

....

....

(iii) In the case where $\theta = \frac{1}{3}\pi$, show that the exact length of the perimeter of triangle ABC is $(24 + 8\sqrt{3})$ cm. [3]

....

....

....

....

....

[Total: 9]

Cambridge International AS & A Level Mathematics, 9709/01 June 2003 Q 9

6 (i) Show that the equation $4\sin^4\theta + 5 = 7\cos^2\theta$ may be written in the form $4x^2 + 7x - 2 = 0$, where $x = \sin^2\theta$. **[1]**

....

....

....

....

....

....

(ii) Hence solve the equation $4\sin^4\theta + 5 = 7\cos^2\theta$ for $0° \leqslant \theta \leqslant 360°$. **[4]**

....

....

....

....

....

....

....

[Total: 5]

Cambridge International AS & A Level Mathematics, 9709/01 November 2002 Q 2

7 In the diagram, OCD is an isosceles triangle with $OC = OD = 10\,$cm and angle $COD = 0.8$ radians. The points A and B, on OC and OD respectively, are joined by an arc of a circle with centre O and radius $6\,$cm.

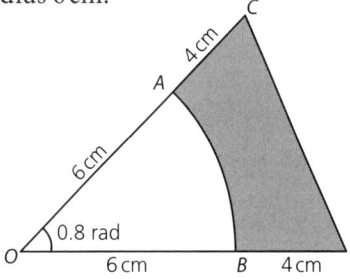

Find

(i) the area of the shaded region, [3]

....

....

....

....

....

(ii) the perimeter of the shaded region. [4]

....

....

....

....

....

[Total: 7]

Cambridge International AS & A Level Mathematics, 9709/01 June 2004 Q 5

8 (i) Sketch and label, on the same diagram, the graphs of $y = 2\sin x$ and $y = \cos 2x$, for the interval $0 \leqslant x \leqslant \pi$. **[4]**

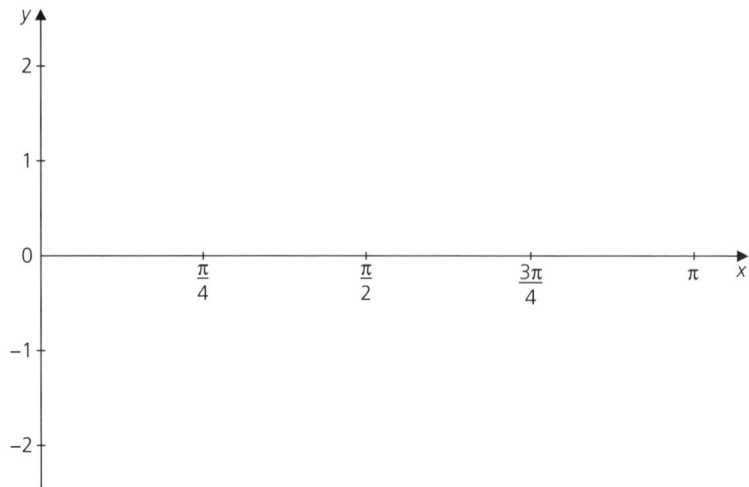

(ii) Hence state the number of solutions of the equation $2\sin x = \cos 2x$ in the interval $0 \leqslant x \leqslant \pi$. **[1]**

....

....

[Total: 5]

Cambridge International AS & A Level Mathematics, 9709/01 November 2004 Q 4

9 The diagram shows a circle with centre O and radius 8 cm. Points A and B lie on the circle. The tangents at A and B meet at the point T, and $AT = BT = 15$ cm.

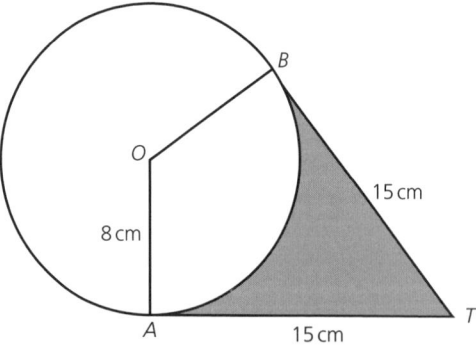

(i) Show that angle AOB is 2.16 radians, correct to 3 significant figures. **[3]**

....

....

....

(ii) Find the perimeter of the shaded region. [2]

....

....

....

(iii) Find the area of the shaded region. [3]

....

....

....

....

....

[Total: 8]

Cambridge International AS & A Level Mathematics, 9709/01 June 2006 Q 7

10 Given that $x = \sin^{-1}\left(\dfrac{2}{5}\right)$, find the exact value of

(i) $\cos^2 x$, [2]

....

....

....

(ii) $\tan^2 x$. [2]

....

....

....

....

[Total: 4]

Cambridge International AS & A Level Mathematics, 9709/01 November 2006 Q 2

11 (i) Show that the equation $3 \sin x \tan x = 8$ can be written as
$3 \cos^2 x + 8 \cos x - 3 = 0$. [3]

....

....

....

....

....

....

....

(ii) Hence solve the equation $3 \sin x \tan x = 8$ for $0° \leqslant x \leqslant 360°$. [3]

....

....

....

....

....

[Total: 6]

Cambridge International AS & A Level Mathematics, 9709/01 November 2007 Q 5

12 The diagram shows a circle with centre O and radius 5 cm. The point P lies on the circle, PT is a tangent to the circle and $PT = 12$ cm. The line OT cuts the circle at the point Q.

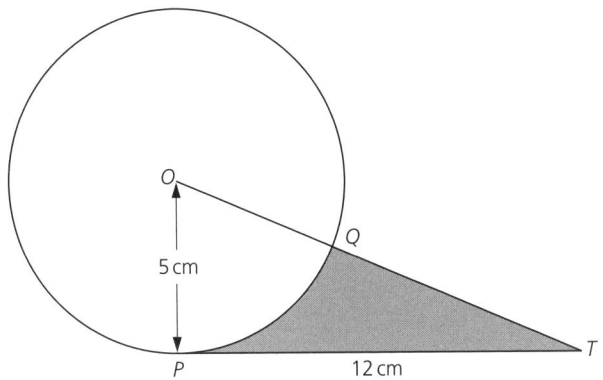

(i) Find the perimeter of the shaded region. [4]

....

....

....

....

....

....

....

....

....

....

(ii) Find the area of the shaded region. [3]

....

....

....

....

[Total: 7]

Cambridge International AS & A Level Mathematics, 9709/01 June 2008 Q 5

13 (i) Show that the equation $2\tan^2\theta\cos\theta = 3$ can be written in the form
$2\cos^2\theta + 3\cos\theta - 2 = 0$. [2]

....

....

....

....

....

....

....

....

(ii) Hence solve the equation $2\tan^2\theta\cos\theta = 3$, for $0° \leqslant \theta \leqslant 360°$. [3]

....

....

....

....

[Total: 5]

Cambridge International AS & A Level Mathematics, 9709/01 June 2008 Q 2

14 The diagram shows a semicircle ABC with centre O and radius 6 cm. The point B
is such that angle BOA is 90° and BD is an arc of a circle with centre A.

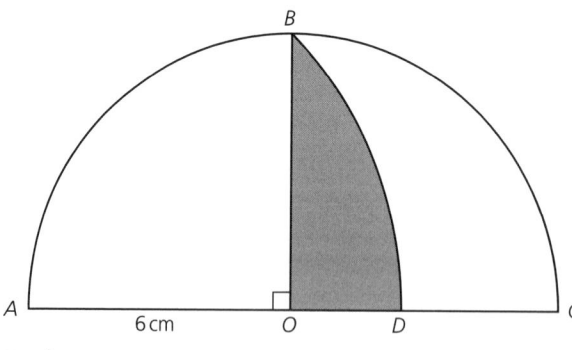

Find

(i) the length of the arc *BD*, [4]

....

....

....

....

(ii) the area of the shaded region. [3]

....

....

....

....

[Total: 7]

Cambridge International AS & A Level Mathematics, 9709/11 November 2009 Q 5

15 The function f is such that $f(x) = 2\sin^2 x - 3\cos^2 x$ for $0 \leqslant x \leqslant \pi$.

(i) Express f(x) in the form $a + b\cos^2 x$, stating the values of *a* and *b*. [2]

....

....

....

....

....

....

(ii) State the greatest and least values of f(x). [2]

....

....

....

....

(iii) Solve the equation f(x) + 1 = 0. [3]

....

....

....

....

....

....

[Total: 7]

Cambridge International AS & A Level Mathematics, 9709/11 June 2010 Q 5

16 Solve the equation $15\sin^2 x = 13 + \cos x$ for $0° \leqslant x \leqslant 180°$. [4]

....

....

....

....

....

....

....

....

Cambridge International AS & A Level Mathematics, 9709/13 November 2010 Q 3

17 (i) Prove the identity $\left(\dfrac{1}{\sin\theta} - \dfrac{1}{\tan\theta}\right)^2 = \dfrac{1-\cos\theta}{1+\cos\theta}$. [3]

....

....

....

....

....

....

....

....

....

(ii) Hence solve the equation $\left(\dfrac{1}{\sin\theta} - \dfrac{1}{\tan\theta}\right)^2 = \dfrac{2}{5}$, for $0° \leqslant \theta \leqslant 360°$. [4]

....

....

....

....

....

....

....

....

[Total: 7]

Cambridge International AS & A Level Mathematics, 9709/13 June 2011 Q 8

18 The diagram shows a circle C_1 touching a circle C_2 at a point X. Circle C_1 has centre A and radius 6 cm, and circle C_2 has centre B and radius 10 cm. Points D and E lie on C_1 and C_2 respectively and DE is parallel to AB. Angle $DAX = \frac{1}{3}\pi$ radians and angle $EBX = \theta$ radians.

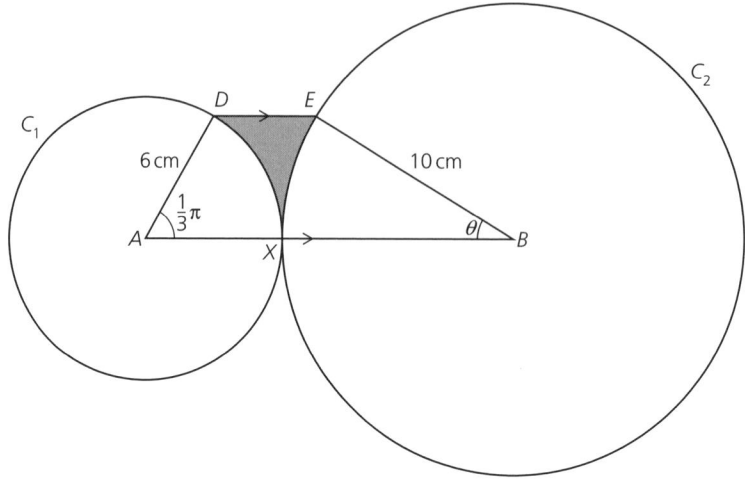

(i) By considering the perpendicular distances of D and E from AB,

show that the exact value of θ is $\sin^{-1}\left(\dfrac{3\sqrt{3}}{10}\right)$. [3]

....

....

....

....

....

....

....

....

....

(ii) Find the perimeter of the shaded region, correct to 4 significant figures. [5]

....

....

....

....

....

....

....

....

....

[Total: 8]

Cambridge International AS & A Level Mathematics, 9709/12 November 2011 Q 6

8 Vectors

1 The diagram shows a solid cylinder standing on a horizontal circular base, centre *O* and radius 4 units. The line *BA* is a diameter and the radius *OC* is at 90° to *OA*. Points *O'*, *A'*, *B'* and *C'* lie on the upper surface of the cylinder such that *OO'*, *AA'*, *BB'* and *CC'* are all vertical and of length 12 units. The mid-point of *BB'* is *M*.

Unit vectors **i**, **j**, and **k** are parallel to *OA*, *OC*, and *OO'* respectively.

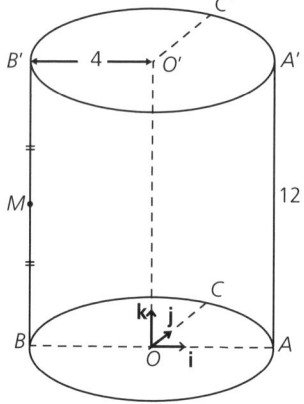

(i) Express each of the vectors $\overrightarrow{MO}$ and $\overrightarrow{MC'}$ in terms of **i**, **j**, and **k**. [3]

....

....

....

....

(ii) Hence find the angle *OMC'*. [4]

....

....

....

....

....

[Total: 7]

Cambridge International AS & A Level Mathematics, 9709/01 June 2002 Q 5

2 The points A, B, C, and D have position vectors $3\mathbf{i} + 2\mathbf{k}$, $2\mathbf{i} - 2\mathbf{j} + 5\mathbf{k}$, $2\mathbf{j} + 7\mathbf{k}$ and $-2\mathbf{i} + 10\mathbf{j} + 7\mathbf{k}$ respectively.

(i)   Use a scalar product to show that BA and BC are perpendicular.   [4]

(ii)   Show that BC and AD are parallel and find the ratio of the length of BC to the length of AD.   [4]

[Total: 8]

Cambridge International AS & A Level Mathematics, 9709/01 June 2003 Q 8

3 Relative to an origin O, the position vectors of the points A, B, C and D are given by

$$\overrightarrow{OA} = \begin{pmatrix} 1 \\ 3 \\ -1 \end{pmatrix}, \quad \overrightarrow{OB} = \begin{pmatrix} 3 \\ -1 \\ 3 \end{pmatrix}, \quad \overrightarrow{OC} = \begin{pmatrix} 4 \\ 2 \\ p \end{pmatrix}, \quad \overrightarrow{OD} = \begin{pmatrix} -1 \\ 0 \\ q \end{pmatrix},$$

where p and q are constants. Find

(i) the unit vector in the direction of $\overrightarrow{AB}$, [3]

(ii) the value of p for which angle $AOC = 90°$, [3]

(iii) the values of q for which the length of $\overrightarrow{AD}$ is 7 units. [4]

[Total: 10]

Cambridge International AS & A Level Mathematics, 9709/01 June 2004 Q 9

4 Relative to an origin O, the position vectors of the points A, B and C are given by

$$\overrightarrow{OA} = \begin{pmatrix} 2 \\ 3 \\ -6 \end{pmatrix}, \quad \overrightarrow{OB} = \begin{pmatrix} 0 \\ -6 \\ 8 \end{pmatrix} \quad \text{and} \quad \overrightarrow{OC} = \begin{pmatrix} -2 \\ 5 \\ -2 \end{pmatrix}.$$

(i) Find angle AOB. [4]

....

....

....

....

....

....

(ii) Find the vector which is in the same direction as $\overrightarrow{AC}$ and has magnitude 30. [3]

....

....

....

....

....

....

(iii) Find the value of the constant p for which $\overrightarrow{OA} + p\overrightarrow{OB}$ is perpendicular to $\overrightarrow{OC}$. [3]

....

....

....

....

....

....

[Total: 10]

Cambridge International AS & A Level Mathematics, 9709/11 November 2009 Q 9

5 The diagram shows the parallelogram $OABC$. Given that $\overrightarrow{OA} = \mathbf{i} + 3\mathbf{j} + 3\mathbf{k}$ and $\overrightarrow{OC} = 3\mathbf{i} - \mathbf{j} + \mathbf{k}$.

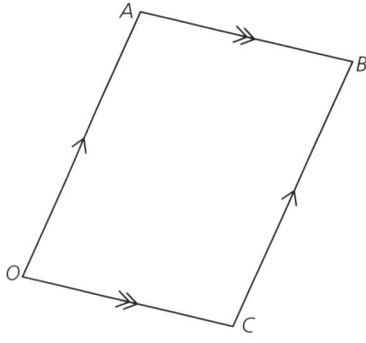

Find

(i) the unit vector in the direction of $\overrightarrow{OB}$, [3]

....

....

....

....

....

(ii) the acute angle between the diagonals of the parallelogram, [5]

....

....

....

....

....

....

....

....

....

(iii) the perimeter of the parallelogram, correct to 1 decimal place. [3]

....

....

....

[Total: 11]

Cambridge International AS & A Level Mathematics, 9709/11 June 2010 Q 10

6 The diagram shows triangle OAB, in which the position vectors of A and B with respect to O are given by $\overrightarrow{OA} = 2\mathbf{i} + \mathbf{j} - 3\mathbf{k}$ and $\overrightarrow{OB} = -3\mathbf{i} + 2\mathbf{j} - 4\mathbf{k}$.

C is a point on OA such that $\overrightarrow{OC} = p\overrightarrow{OA}$, where p is a constant.

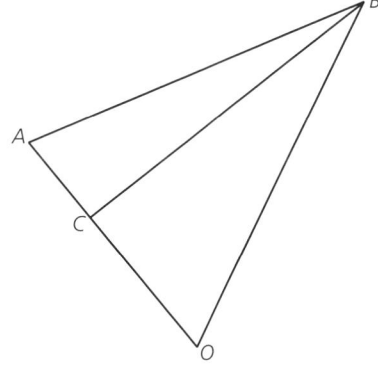

(i) Find angle AOB. [4]

....

....

....

....

....

....

....

(ii) Find $\overrightarrow{BC}$ in terms of p and vectors **i**, **j** and **k**. [1]

....

....

....

....

....

....

....

(iii) Find the value of p given that BC is perpendicular to OA. [4]

....

....

....

....

....

....

....

....

[Total: 9]

Cambridge International AS & A Level Mathematics, 9709/13 November 2010 Q 10

7 In the diagram, *OABCDEFG* is a rectangular block in which $OA = OD = 6$ cm and $AB = 12$ cm. The unit vectors **i**, **j** and **k** are parallel to $\overrightarrow{OA}$, $\overrightarrow{OC}$ and $\overrightarrow{OD}$ respectively. The point *P* is the mid-point of *DG*, *Q* is the centre of the square face *CBFG* and *R* lies on *AB* such that $AR = 4$ cm.

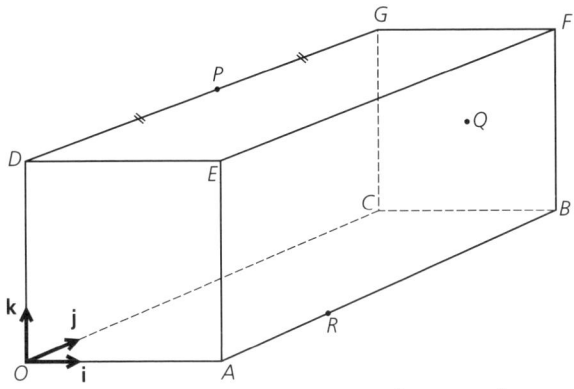

(i) Express each of the vectors $\overrightarrow{PQ}$ and $\overrightarrow{RQ}$ in terms of **i**, **j** and **k**.　　　　　　[3]

....

....

....

(ii) Use a scalar product to find angle *RQP*.　　　　　　[4]

....

....

....

....

....

....

[Total: 7]

Cambridge International AS & A Level Mathematics, 9709/13 June 2011 Q 5

Formula sheet

PURE MATHEMETICS

Algebra

For the quadratic equation $ax^2 + bx + c = 0$:

$$x = \frac{-b \pm \sqrt{(b^2 - 4ac)}}{2a}$$

For an arithmetic series:

$$u_n = a + (n-1)d, \qquad\qquad S_n = \frac{1}{2}n(a+l) = \frac{1}{2}n\{2a + (n-1)d\}$$

For a geometric series:

$$u_n = ar^{n-1}, \quad S_n = \frac{a(1-r^n)}{1-r} \ (r \neq 1), \quad S_\infty = \frac{a}{1-r} \ (|r| < 1)$$

Binomial expansion:

$$(a+b)^n = a^n + \binom{n}{1}a^{n-1}b + \binom{n}{2}a^{n-2}b^2 + \binom{n}{2}a^{n-3}b^3 + ... + b^n, \ \text{ where } n \text{ is positive}$$

integer and $\binom{n}{r} = \dfrac{n!}{r!(n-r)!}$

Trignometry

Arc length of circle $= r\theta$ (θ in radians)

Area of sector of circle $= \frac{1}{2}r^2\theta$ (θ in radians)

$$\tan\theta \equiv \frac{\sin\theta}{\cos\theta}$$

$$\cos^2\theta + \sin^2\theta = 1$$

Differentiation

$f(x)$	$f'(x)$
x^n	nx^{n-1}

Integration

$f(x)$	$\int f(x)\,dx$
x^n	$\dfrac{x^{n+1}}{n+1} + c \ (n \neq -1)$

Vectors

If $\mathbf{a} = a_1\mathbf{i} + a_2\mathbf{j} + a_3\mathbf{k}$ and $\mathbf{b} = b_1\mathbf{i} + b_2\mathbf{j} + b_3\mathbf{k}$ then

$\mathbf{a}.\mathbf{b} = a_1b_1 + a_2b_2 + a_3b_3 = |\mathbf{a}||\mathbf{b}|\cos\theta$

Principal values:

$-\dfrac{1}{2}\pi \leq \sin^{-1}x \leq \dfrac{1}{2}\pi$

$0 \leq \cos^{-1}x \leq \pi$

$-\dfrac{1}{2}\pi < \tan^{-1}x < \dfrac{1}{2}\pi$